T0249594

Monolithic Nanoscale Photonics—Electronics Integration in Silicon and Other Group IV Elements

Monolithic Nanoscale Photonics—Electronics Integration in Silicon and Other Group IV Elements

Henry Radamson

Lars Thylén

AMSTERDAM • BOSTON • HEIDELBERG • LONDON
NEW YORK • OXFORD • PARIS • SAN DIEGO
SAN FRANCISCO • SINGAPORE • SYDNEY • TOKYO

Academic Press is an imprint of Elsevier

Academic Press is an imprint of Elsevier
32 Jamestown Road, London NW1 7BY, UK
525 B Street, Suite 1800, San Diego, CA 92101-4495, USA
225 Wyman Street, Waltham, MA 02451, USA
The Boulevard, Langford Lane, Kidlington, Oxford OX5 1GB, UK

Copyright © 2015 Elsevier LTD. All rights reserved.

No part of this publication may be reproduced or transmitted in any form or by any means, electronic or mechanical, including photocopying, recording, or any information storage and retrieval system, without permission in writing from the publisher. Details on how to seek permission, further information about the Publisher's permissions policies and our arrangement with organizations such as the Copyright Clearance Center and the Copyright Licensing Agency, can be found at our website: www.elsevier.com/permissions

This book and the individual contributions contained in it are protected under copyright by the Publisher (other than as may be noted herein).

Notices
Knowledge and best practice in this field are constantly changing. As new research and experience broaden our understanding, changes in research methods, professional practices, or medical treatment may become necessary.

Practitioners and researchers must always rely on their own experience and knowledge in evaluating and using any information, methods, compounds, or experiments described herein. In using such information or methods they should be mindful of their own safety and the safety of others, including parties for whom they have a professional responsibility.

To the fullest extent of the law, neither the Publisher nor the authors, contributors, or editors, assume any liability for any injury and/or damage to persons or property as a matter of products liability, negligence or otherwise, or from any use or operation of any methods, products, instructions, or ideas contained in the material herein.

British Library Cataloguing-in-Publication Data
A catalogue record for this book is available from the British Library

Library of Congress Cataloging-in-Publication Data
A catalog record for this book is available from the Library of Congress

ISBN: 978-0-12-419975-0

For information on all Academic Press publications
visit our website at **http://store.elsevier.com/**

Working together
to grow libraries in
developing countries

www.elsevier.com • www.bookaid.org

CONTENTS

ACKNOWLEDGMENTS

The authors acknowledge valuable discussions with and material from Prof. Sebastian Lourdudoss, Associate Prof. Lech Wosinski, Prof. Hans Ågren, Prof. Anders Hult, and Mr. Mahdi Moeen, Mr. Milad Ghadami all from the Royal Institute of Technology (KTH).

Everybody is in some way or another acquainted with or affected by the enormous impact of integrated electronics and integrated circuits (ICs). They have shaped virtually all aspects of our social, professional, and cultural lives. Their development saw its beginning in the 1960s with the emergence of ICs, awarded with the Nobel Prize in 2000. The material of choice has been mainly silicon which exists in abundance on the earth and its development is hugely aided by the existence of a natural passivating oxide, protecting the crystalline circuits.

At the same time, in the 1960s and maybe inspired by the unfolding of integrated electronics, another concept saw the light at the famed Bell Labs: integrated optics (today named integrated photonics, involving photonics integrated circuits, PICs). Though superficially related, they were in fact very different. Somewhat simplifying, we are physically dealing with fermions in electronics and bosons in photonics. Thus, in photonics, one was over the years working with several various constituent materials and device structures, in contrast to ICs.

Photonics has over the last decades developed into a key enabling technology, with inroads in information and communications technology (ICT) with the optical fiber as one of the landmarks, awarded with the Nobel Prize in 2009; in biotechnology with a wealth of sensors and in lighting and energy. All these fields are more or less amenable to integrated photonics. In addition, we have areas such as medicine (for therapy and diagnosis), manufacturing (e.g., high-power fiber lasers), security, and surveillance. This shows the tremendous versatility and impact of photonics and makes photonics a counterpart to electronics, with different but complementary functions.

But there are also basic physical differences between *electronics* in the shape of electronic ICs and ancillary devices and *photonics*: As mentioned, in the former case, we are generally dealing with fermions (electrons), which obey Fermi–Dirac statistics, whereas in the latter case we are employing bosons (photons), obeying Bose–Einstein statistics. This has significant but not immediately obvious consequences in the sense that it appears all but possible to create photonics devices that perform *digital signal processing* and *RAM-type memory functions*, operations in which electronics excel. On the other hand, photonics is the technology of choice for high-speed transmission and routing of vast amounts of data, necessary in our information society built on the Internet. This unique capability of photonics is based on the huge optical frequency, 300 THz at a wavelength of 1 μm, with a concomitant information bandwidth about three orders of magnitude larger than that of electronics. But it should be noted that photonics, as described in this book, has a wider applications envelope than ICT, as noted above.

This book describes the basics of integrated electronics and integrated photonics, notably in a language understandable for readers at the EE masters or BSc levels or equivalent. Thus, the goal is to create a cross-disciplinary understanding in important and developing fields, with large potential ramifications and encourage cooperation over discipline boundaries.

The book has a focus on group IV elements, notably silicon: These materials are totally dominating in electronics since a long time ago and recently silicon has also come into focus for integrated photonics, with concomitant prospects of merging the two technologies into something appropriately named integrated electronics–photonics. It is anticipated that such a synergetic combination of integrated electronics and integrated photonics, in concepts not yet materialized or invented, will have a very large impact.

We live in a macro world which is being controlled by nano-properties.

CHAPTER 1

Metal Oxide Semiconductor Field Effect Transistors

Monolithic Nanoscale Photonics—Electronics Integration in Silicon and Other Group IV Elements.
DOI: http://dx.doi.org/10.1016/B978-0-12-419975-0.00001-5
© 2015 Elsevier Ltd. All rights reserved.

PART ONE: BASICS OF METAL OXIDE SEMICONDUCTOR FIELD EFFECT TRANSISTORS

The **metal oxide semiconductor field effect transistor** (**MOSFET**) is the most basic element of integrated circuits (IC) which is used in many electrical devices in our daily life. The transistor has four-terminal contacts: gate (G), source (S), drain (D), and substrate body (B). In a transistor, the body terminal can be connected to the source to make it a three-terminal device. MOSFETs have different symbols depending on the operating mode and type of dopant in the channel as shown in Figure 1.1. The drive carriers in nMOS is electrons (n-channel) but the body is p-type doped meanwhile pMOS the main carriers are holes (p-channel) with n-type doped body

In principle, MOSFET structure consists of two PN junctions (SB and DB regions) and carrier transport occurs in channel region between source and drain. The SB and DB PN junctions are connected back-to-back and this prevents the current flow in the channel. By applying a voltage to the gate contact, an electric field is established through the oxide and the transistor can be turned on or off. MOSFETs can be used as switch or amplifier in the circuits.

In case of a switch, the resistance R_{DS} is dependent on V_{GS} ($R_{DS} = f(V_{GS})$). The transistor is in on-state when a large on-state current (I_{on} is high) flows and it is off when a very small off-state leakage current is obtained (I_{off} is low). A MOSFET acts as an amplifier when $I_D = f(V_{GS})$ or the drain current is voltage-controlled.

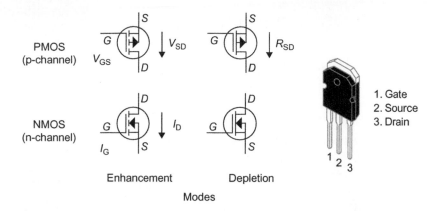

Figure 1.1 FET's symbols with related voltages (V$_{SD}$ and V$_{GS}$), currents (I$_D$ and I$_G$), and resistance (R$_{SD}$).

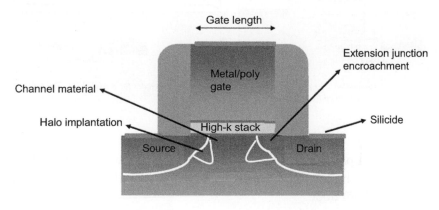

Figure 1.2 A schematic of MOSFET structure.

MOSFETs are manufactured in 2D or 3D designs on silicon bulk or silicon on insulator (SOI) substrates. The traditional MOSFET source/drain (S/D) junctions and the extended junctions are formed by implantation. A tilted angle implantation is also used to further adjust the doping profile below the gate region (so-called Halo implantation). Stressor materials such as silicon−germanium, carbon-doped silicon, or germanium−tin may induce strain in the channel region to enhance the carrier mobility. Nitride layers which are deposited over the transistor structure are also used as stressor materials. The channel material is Si, but other novel materials such as germanium, germanium−tin, III−V materials, graphene, and graphene-like materials have been used too. The dielectric material was for a long time SiO_2 (until 90 nm node) which later replaced by high-*k* materials to avoid the leakage of thin oxide layers. In order to decrease the contact resistances in the transistors silicide layers were formed prior to meatllization. Figure 1.2 shows a 2D MOSFET structure

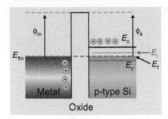

Figure 1.3 Energy band diagram for metal–oxide–semiconductor.

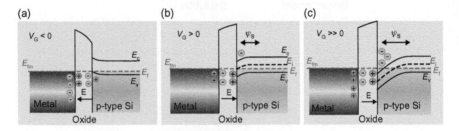

Figure 1.4 (a) Accumulation, (b) depletion, and (c) inversion operation modes for a MOSFET. The mid bandgap and fermi levels are marked by E_i *and* E_f.

where different parts of a transistor are indicated. Sidewall transfer lithography (STL) technique has been widely used to manufacture MOSFETs in down to 14 nm node.

SURFACE SPACE–CHARGE REGIONS IN MOSFETs

The MOSFET structure has three main parts: metal, oxide, and semiconductor. Initially, the Fermi level is aligned and the energy bands are flat and the workfunctions are the same for all three parts as shown in Figure 1.3. In this situation, there is no charge neither inside the oxide nor at the interface between the oxide and semiconductor.

Applying a voltage to the gate electrode causes an electric field E between the both sides. This makes a displacement of carriers near each side forming a two space–charge region.

When no charge is present at the interface between the oxide and semiconductor, the flatband voltage (V_{FB}) is described as difference between the gate metal workfunction (φ_M) and the semiconductor workfunction, ϕ_S: $V_{FB} = \varphi_M + \phi_S$.

The charge density in the semiconductor (channel region) is dependent on the applied voltage to the metal (or poly) gate. Figure 1.4(a)–(c) shows three conditions for an nMOSFET that can occur when a voltage is applied to the gate terminal: accumulation, depletion, and inversion.

An nMOSFET is in on-state when a negative voltage is applied to the gate and the conductivity of Si channel is increased. The negative voltage (φ_s) results in a bending of the energy bands upward (E_f is constant). The modified energy band makes more holes accumulated in channel region as shown in Figure 1.4(a). In this case, the gate voltage is distributed across the oxide layer and the formed voltage in semiconductor is given by $V_{GS} = V_{ox} + \psi_{Si}$.

In general, the density of accumulated carriers at the surface for a biased gate can be obtained as the following:

$$n = n_i e^{\frac{(E_f - E_i)}{kT}} \tag{1.1a}$$

$$p = p_i e^{\frac{(E_i - E_f)}{kT}} \tag{1.1b}$$

where E_i is the mid-gap energy in the bandgap. More carriers will be available when the band-bending increases. These equations show that for the carrier accumulation condition in an nMOSFET, $p > n$ when $E_i - E_f$ becomes large (holes in the channel region).

The carrier condition changes when a small or moderate positive voltage is applied to the gate. This is known as carrier depletion for the transistors and at this time the energy bands bend downward as shown in Figure 1.4(b). The value of $E_i - E_f$ decreases and the channel region is now depleted of holes. The total charge in the established space–charge region in semiconductor with the depletion width (t_d) is obtained from the equation

$$Q_s = q \, N_s \, t_d \tag{1.2}$$

where N_s is the dopant concentration in the semiconductor. The negative sign is due to the polarity of the charge in the depletion region. The depletion length in the semiconductor is obtained by solving the Poisson equation. Therefore, the formed electric field, ξ_s in semiconductor's depletion region can be written as

$$\frac{d\xi}{dx} = \frac{qN_s}{K_s \varepsilon_0} \quad (0 < x < t_d) \Rightarrow \xi(x) = \frac{qN_s}{K_\varepsilon \varepsilon_0}(t_d - x) \tag{1.3}$$

where K_s is dielectric constant of semiconductor and ε_0 is permittivity. The electric field in Eq. (1.3) is defined in terms of the raised electrostatic potential as $\xi(x) = -d\psi_s/dx$ and after integration, the potential change is written as:

$$\psi_s(t_d) - \psi_s(x) = \frac{qN_s}{2K_\varepsilon\varepsilon_0}(t_d - x)^2 \qquad (1.4)$$

In case of $x = 0$, the potential denoted by ψ_s has the following relationship with the depletion width:

$$\psi_s = \frac{qN_s t_d^2}{2K_s\varepsilon_0} \qquad (1.5)$$

If a relatively large positive voltage is applied to the gate electrode, then the band-bending downward becomes severe. This condition may cause the mid-gap energy E_i to pass E_f near the Si and oxide interface. As a consequence, an inversion layer is established where the electrons will have greater density than holes. A strong inversion layer is desired for MOSFETs operation, as shown in Figure 1.4(c). When the density of electrons becomes equal the to the hole density in the bulk, a moderate inversion has been achieved.

Threshold voltage (V_T) is a critical voltage for MOSFETs and it refers to the situation when the applied gate voltage (V_{GS}) is reached to a value resulting in a significant extended inversion layer and rapid inverse charge. Thus the voltages can be written from flatband voltage equation as

$$V_T = -\frac{Q}{C_0} + \phi_{Si} \quad \text{where} \quad C_0 = \frac{K_0\varepsilon_0}{x_0} \qquad (1.6a)$$

We notify here that a band bending occurs in the Si side due to workfunction differences between the Si and metal, charges in oxide and surface states.

Then, the threshold voltage is modified to:

$$V_T = V_{FB} + \phi_{Si} - \frac{Q_s}{C_0} \qquad (1.6b)$$

where φ_{Si} is the workfunction of Si. In eq. (1.6b), Q_s under strong inversion can be obtained from $Q_s = -\sqrt{2qK_s\varepsilon_0 N_a(\psi_s + \varphi_{Si})}$ where ψ_s is the voltage of the n-inversion side of the junction where Si is p-type doped.

LEAKAGE COMPONENTS IN MOSFETs

There are different leakage components in MOSFETs depending on the transistor size: subthreshold leakage current, gate–oxide leakage, and junction leakage.

These leakage sources have become more important due to the continuous downscaling of transistors. The supply voltage is being decreased to reduce both dynamic power consumption and electric fields inside the transistor structure. All leakage sources in a transistor manufactured in a technology generation may cause in the dissipation of a large amount of power.

SUBTHRESHOLD CURRENT

From the definition of threshold voltage, it is concluded that the inversion layer charge is zero below this voltage. But in fact, there is a subthreshold current which decreases exponentially by gate voltage decrease below the threshold voltage as shown in Figure 1.5.

The amount of subthreshold drive current is regulated by the threshold voltage which lies between ground and the supply voltage. The gate voltage is expressed in terms of semiconductor potential and flatband voltage as the following:

$$V_G = V_{FB} + \psi_s + V_o = V_{FB} + \psi_s + \frac{\sqrt{2q\varepsilon_s \psi_s}}{C_0} \qquad (1.7)$$

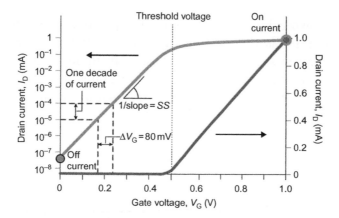

Figure 1.5 I (V) curve of a MOSFET when the gate voltage of typically 1 V is applied. The subthreshold voltage is estimated to 80 mV in the figure [1].

The variation of the gate voltage in terms of semiconductor potential is expressed as

$$\frac{dV_G}{d\psi_s} = 1 + \frac{1}{2C_0}\sqrt{\frac{q\varepsilon_s}{\psi_s}} \approx 1 + \frac{1}{2C_0}\sqrt{\frac{q\varepsilon_s}{2\phi_f}} = n \qquad (1.8)$$

where the semiconductor potential is estimated to be $2\phi_f$ for threshold. Then, subthreshold current is obtained from

$$I_D \propto Q \propto \exp\left(\frac{\psi_s}{V_T}\right) \propto \exp\left(\frac{V_G}{nV_T}\right) \qquad (1.9)$$

GATE—OXIDE LEAKAGE

The oxide breakdown and oxide reliability are issues arisen by downscaling of the MOSFETs. The tunneling of carriers through the oxide is enhanced when its thickness is shrunk. The carriers gradually damage the oxide layer which is known as time-dependent destructive breakdown (TDDB). High-k materials with large dielectric constant may solve this problem. Typically HfO, Al_2O_3, and ZrO dielectrics have been integrated in MOSFETs [2]. A wider overlook of high-k materials is provided in Chapter 4.

S/D JUNCTION LEAKAGE

Both source-to-substrate and drain-to-substrate junctions can affect many transistor characteristics: the current—voltage characteristics, lowering output resistance, and the speed of transistor. The junction leakage has influence on the standby power dissipation.

Due to downscaling of MOSFETs, the junctions are designed for higher doping levels (halo doping is applied) and shallower depths. Heavier doping in source and drain results in thinner depletion layers and more recombination centers. An annealing step is applied to remove the lattice damage after junction dopant implantation, which can further increase the junction leakage.

MOS CAPACITORS

The MOS structures in Figure 1.3 are basically capacitors with the oxide (SiO_2 or high-k oxide) as dielectric material. The capacitances of space—charge layer in Si (C_s) and oxide (C_{ox}) with thickness of x_0 are in series and are defined as

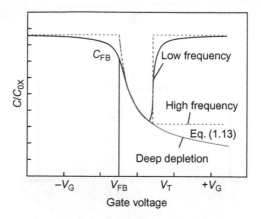

Figure 1.6 C–V characteristic for low and high frequencies [3].

$$C_s = \frac{K_s \varepsilon_0}{x_d} \tag{1.10a}$$

$$C_{ox} = \frac{K_s \varepsilon_0}{x_0} \tag{1.10b}$$

The gate voltage is expressed as

$$V_G = -\frac{Q_s}{C_s} + \psi_s \tag{1.11}$$

Then the depletion thickness will be

$$t_d = \frac{K_s \varepsilon_0}{C_{ox}} \left(\sqrt{1 + \frac{2V_G C_{ox}^2}{q K_s \varepsilon_0 N_a}} - 1 \right) \tag{1.12}$$

The total capacitance, C $(C = C_s C_{ox}/(C_s + C_{ox}))$ for MOS structure is easily obtained from the following equation (as shown in Figure 1.6):

$$C = \frac{C_{ox}}{\sqrt{1 + (2C_{ox}^2/q N_a K_s \varepsilon_0)V_G}} \tag{1.13}$$

STATIC CHARACTERIZATION OF MOSFETs

In order to discuss the I–V characteristics of MOSFETs, different parameters have to be defined:

Threshold voltage (V_T).
Voltages between D and S terminals (V_{DS}) and between G and S terminals (V_{GS}).

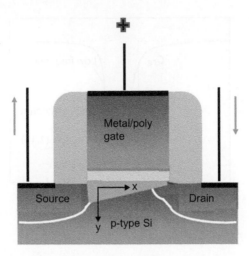

Figure 1.7 A schematic of a biased nMOSFET. The marked triangle shows the formed potential in the channel region.

Supply voltage (V_{DD}).
Overdrive voltage (V_{ov}): when V_{GS} exceeds the threshold. V_{OV} is defined as $V_{GS} - V_{T}$.
Channel length and width, L and w (aspect ratio: L/W).
Oxide capacitance (C_{ox}).
Channel mobility (μ_e).

Coefficient of conductance (k_n) is defined as $k_n = \mu_e C_{ox}(W/L)$

A schematic of an nMOSFET structure is shown in Figure 1.7. When no voltage is applied to the gate terminal, the conductance in the channel is low and no current flows between source and drain terminals ($I_D = 0$). This situation changes rapidly if a positive voltage is applied to the gate. In order to have a large enough conductance in the channel so that the electrons flow, the gate applied voltage has to be greater than the threshold voltage ($V_{GS} > V_T$).

Upon flow of current between source and drain terminals with assumption of $V_{GS} > V_T$, a graded voltage forms across the channel region as shown in Figure 1.8. It is important that $V_{DS} < V_{GS} - V_T$, otherwise a pinch through situation may occur.

In this case, V_{GS} will control the conductance of the channel or, in other words, the channel acts as a variable resistor. Thus, the

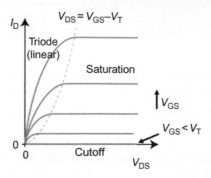

Figure 1.8 IV characteristics of an n-channel MOSFET in different operation modes [3].

conductance between source and drain (g_{DS}) is dependent on both V_{GS} and V_{DS}. The g_{DS} is obtained from this expression:

$$g_{DS} = \frac{1}{r_{DS}} = k_n \left(V_{GS} - V_t - \frac{1}{2} V_{DS} \right) \tag{1.14}$$

where r_{DS} is the resistance between source/drain.

The induced channel charge is written as the following: $Q_I = - C_{ox}(V_G - V_T - \psi)$.

Therefore, the channel current can also be calculated by carrier mobility, the formed electric field along the channel direction, channel length, and channel charge according to

$$I_D = W \mu_n Q_I \xi_y \tag{1.15}$$

where the formed electric field along the channel direction is given by: $\xi_y = -d\psi/dy$

Therefore, I_D relationship can be rewritten as

$$I_D \, dy = W \mu_n C_{ox}(V_{GS} - V_T - \psi)d\psi \Rightarrow$$

$$I_D = \mu_n C_{ox} \frac{W}{L} \left[(V_{GS} - V_T)V_{DS} - \frac{1}{2} V_{DS}^2 \right] \tag{1.16a}$$

Then, the expression may be rephrased using coefficient of conductance as

$$I_D = k_n \left[(V_{GS} - V_T)V_{DS} - \frac{1}{2}V_{DS}^2 \right] \tag{1.16b}$$

The I_D expression is a quadratic function of V_{DS} with a maximum point at V_T. Expression 1.15 is very important for transistor characterization and is applied to derive the channel mobility from the electrical measurements.

An nMOSFET will have three modes with the following condition as shown in Figure 1.8:

1. Cutoff: when $V_{GS} < V_T$ and the channel current $I_D = 0$ A.
2. Triode or linear mode if $0 < V_{DS} < V_{GS} - V_T$ conductance is high for the carriers to flow along the channel with a current of I_D. There are two electrical fields along x- and y-directions that are marked in Figure 1.7. The electrical field along y generates the inversion layer, whereas the field along x-direction causes the carriers move along the channel direction.
3. Saturation mode when $V_{DS} \geq V_{GS} - V_T$. This is a voltage-controlled current source (VCCS). The drain current can be rewritten as $I_D = (1/2)k_n(V_{GS} - V_T)^2$. So far, the characteristics of nMOS have been discussed but for pMOS all currents and voltages and voltages conditions will be reversed. This means that V_T, V_{SG}, V_{DS} and V_{OV} will be negative and for example, conditions for saturation mode $V_{GS} < V_T$ and $V_{DS} < V_{OV}$.

The MOSFET operates as a switch when it shifts between linear and cutoff modes. The transistor may act as an amplifier in the circuit as shown in Figure 1.9.

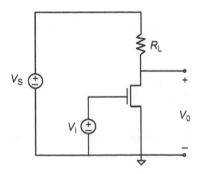

Figure 1.9 An nMOS transistor in a circuit acting as an amplifier.

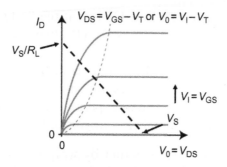

Figure 1.10 IV characteristics of an nMOSFET designed as an amplifier.

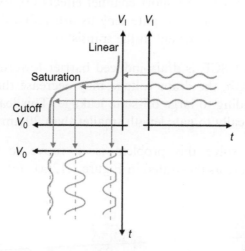

Figure 1.11 Signals to n-channel MOSFET in different operation modes.

In this simple circuit, R_L may act as a load resistance to the transistor. The transistor acts as an amplifier at the intersections of load line with the MOSFET characteristic curves as shown in Figure 1.10.

When a signal is delivered to a MOSFET, the outcome can be illustrated for different modes as shown in Figure 1.11.

TRANSFER FROM 2D TO 3D NANOSCALED TRANSISTORS

Over four decades, MOSFET was manufactured in 2D structure until a nonplanar 3D transistor was introduced in the form of a fin field effect transistor (FinFET) processed on SOI substrates. The term FinFET was introduced by University of California [4]. FinFETs have entirely different design and characteristics than the planar transistors. The distinctive characteristic of a FinFET is that the body of the device is a thin fin of Si where its thickness is the effective channel length of the transistor. Therefore, all categories of these transistors are also known as multigate FETs (MUGFETs). In general, most of the MUGFETs have a number of parallel nanowires (the so-called fingers) which have a joint gate electrode. This structure permits the current drive to be enlarged by increasing the number of fingers of the transistor.

A benefit of 3D transistors is that by wrapping around the transistor gate, a better control over electrical transport through the channel is provided. Therefore, the short channel effects (SCEs) which are an important problem for small gate lengths are reduced, resulting in a decrease of the leakage current in the transistor.

An example of SCE is drain-induced barrier lowering (DIBL). For a small gate length, the drain voltage may decrease the barrier height of the source leading to large off-state leakage. This leakage phenomenon limits practically to gate length smaller than 20 nm [5].

One way to solve this problem is to move to multigate fully depleted transistors as illustrated in Figure 1.12(a)–(e).

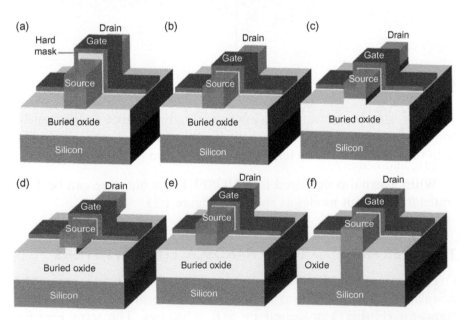

Figure 1.12a–f Schematic of 3D transistor structures: (a) SOI FinFET, (b) SOI triple-gate (or tri-gate) MOSFET, (c) SOI Π-gate MOSFET, (d) SOI Ω-gate MOSFET, (e) SOI gate-all-around MOSFET, and (f) A bulk tri-gate MOSFET. [1].

Triple-gate or tri-gate MOSFET is a 3D design for transistors which Intel introduced for 22 nm node in 2011. The most important advantage of 3D tri-gate transistors is that they operate at lower voltage decreasing significantly the active power.

Tri-gate FETs differ slightly from FinFETs. FinFET has a block material in the form of a thick dielectric on the top and is considered as a dual-gate transistor in the device terminology. The block material is introduced to avoid the forming of an inversion channel at the top part of the Si-fin (see Figure 1.12(a)). Meanwhile, the gate control in tri-gate FETs is employed on the channel region from all three sides of the device (see Figure 1.12(b)).

The gate control can be further increased compared to tri-gate by introducing a bottom side in Π-gate MOSFET (Figure 1.12(c)). In Ω-gate MOSFET, the gate control from the bottom side of the channel is improved compared to Π-gate MOSFET (Figure 1.12(d)). Both Π- and Ω-gate MOSFETs are named after the shape of their gates. If the gate is formed all around the channel, an all-around MOSFET is manufactured (Figure 1.12(e)). When a tri-gate MOSFET is formed on a bulk substrate instead of a SOI wafer, the transistor is called a bulk tri-gate MOSFET (Figure 1.12(f)).

In this group of transistors, bulk FinFET is considered as a cheaper alternative to SOI FinFET. Although the cost of the substrate is lower for bulk Si wafers compared to SOI, the fin formation needs six to eight extra process steps [6].

FinFETs may demonstrate different electrical behavior depending on how they are manufactured. For example, the dimension (width, length, and height) and shape (rectangular, trapezoidal, and triangular) of Si-fin as well as body doping level will affect the threshold voltage and carrier profile of FinFET. Sidewalls with more inclined angle and a low doping level of $\sim 10^{17} \, cm^{-3}$ are required to increase the threshold voltage [7].

GATE INTEGRATION IN FinFETs

One of the issues which appears for the nanoscaled transistors is the gate formation and the integration of the high-k material and metal gate within the thermal budget of the device process (see more

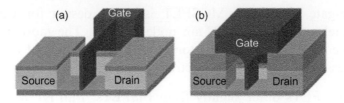

Figure 1.13 The FinFET structures with integrated (a) gate-first and (b) gate-last processes.

discussions in chapter 4). The high-k materials are amorphous and a high-temperature annealing treatment is required for their crystallization. Then either high thermally stable materials are sought or the gate should be processed last. Many research studies have compared the benefits and drawbacks of gate-first and gate-last processes for transistors (Figure 1.13).

For the gate-last process, a dummy gate is primarily processed by using SiO_2 before the selective epitaxy of SiGe in source and drain regions. Later in the process, the dummy gate is entirely removed and the high-k material is deposited in the gate region.

PARASITIC SOURCES IN MOSFET STRUCTURE

The main factors which put an end to the downscaling of planar transistors after 40 years were the parasitic capacitances and resistances. These limiting factors are eliminated significantly by transition to 3D transistors. The origin of these parasitic effects is due to the small space between neighboring devices (only few nanometers) and the shallow S/D junctions where the contact size has been truly scaled to maintain the high density in the absence of gate length scaling.

The small contact size results in higher contact resistance as well as contact-to-gate capacitance. Traditionally, the parasitic effects had no large impact on the transistor performance because they were significantly smaller than the resistance and capacitance of the channel. However, both these effects are proportional to the gate length which has been significantly reduced during the past years.

In other words, the parasitic effects became comparable or even larger than the intrinsic channel (gate and body) capacitance and resistance (see Figure 1.14(a) and (b)).

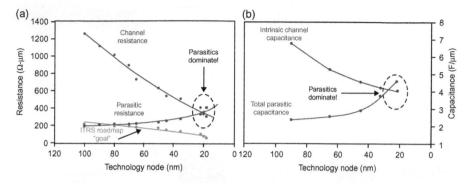

Figure 1.14 Parasitic resistances and capacitances in different technology nodes.

One of the most important parts of today's research is finding new channel materials with high carrier mobility. The parasitic resistances and capacitances are still the main obstacle for integration. As an example, carbon nanotube transistors show channel mobility of a few orders of magnitude higher than Si; however, due to the presence of parasitic resistances, the manufactured inverters are significantly slower than the state-of-the-art MOSFETS [8,9]. Another example that can be pointed out here is multigate transistors consisted of Si-fins, where the parasitic resistance to small S/D areas is higher than planar MOSFETs [10].

LITHOGRAPHY OF NANOSCALED MOSFETs

Various lithography techniques have been reviewed in Chapter 4. Here the focus is on Sidewall Transfer Lithography (STL), which is a simple and cost-effective technique to create nanoscaled transistors.

SIDEWALL TRANSFER LITHOGRAPHY

To follow the Moore's law, it is required that the number of transistors in the chip is almost doubled every 2 years. This requires a downscaling of MOSFETs both in vertical and horizontal dimensions. The gate length of MOSFETs should be thinned and therefore lithography technique to achieve the nanoscale levels becomes very critical. One concern with submicron gate lengths in planar MOSFETs or 3D FinFETs is line-edge roughness (LER) [11]. This is caused by polymer aggregates which naturally exist in photoresist films [12]. For nanoscaled MOSFETs, the LER problem results in a large line-width roughness (LWR) which degrades device performance. The STL is a method

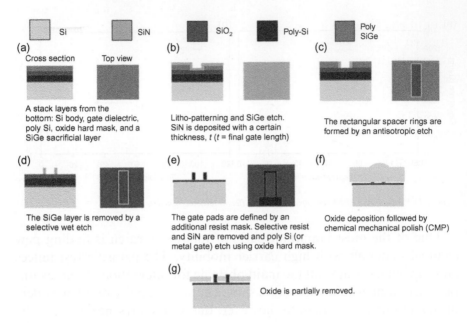

Figure 1.15 STL process steps for FinFET fabrication.

which provides an excellent opportunity to achieve the nanoscaled dimensions without expensive lithography techniques. The STL technique is performed by a deposition and etching of sacrificial layers. Figure 1.15(a)–(g) shows the whole process to form nanowires. The main idea is deposition of a SiN layer with a thickness of the desired gate length of the transistor. The SiN is dry-etched and leaving behind SiN sidewall spacers which can be used later as hard mask for the gate formation [13].

STL technique has been successfully applied to form regular shape Si-fin with 14 nm width. A SiGe layer is deposited selectively on Si-fins using reduced pressure chemical vapor deposition (RPCVD) technique.

PART TWO: STRAIN ENGINEERING IN GROUP IV MATERIALS

Strain is a mechanical distortion caused by an exerted force to the semiconductor crystal. As a result, the electrical, mechanical, and optical properties of the semiconductor will change. The acting force is caused either by an epitaxial layer or by processing method when a kind of stressor material is deposited on the top or on both sides of the strained layer. The direction and dimension of the exerted force is important to determine the effect of the strain. If a crystal cubic is compressed inward in x- and y-directions, then the cubic becomes elongated in z-direction and the strain is called compressive. In contrary, if the cubic is exposed to an outward force in x- and y-directions, then the crystal is shrunk in z-direction, and the strain is called tensile.

Group IV materials have diamond crystal structure and the strain changes the cube into a tetragonal form. Examples for compressive- and tensile-strained materials could be $Si_{1-x}Ge_x$ on Si (or $Ge_{1-x}Sn_x$ on strain-relaxed Ge) and $Si_{1-y}C_y$ on Si (or Ge on strain-relaxed $Ge_{1-x}Sn_x$), respectively. The strain in a layer resolved into vertical ($\varepsilon_\perp$) and in-plane ($\varepsilon_{//}$) components are defined according to the formulas:

$$\varepsilon_\perp = \frac{a_\perp - a_b}{a_b} \quad \text{and} \quad \varepsilon_{//} = \frac{a_{//} - a_b}{a_b} \qquad (1.17a \text{ and } 1.17b)$$

where the lattice constants are marked in Figure 1.16. The components of compressive strain are $\varepsilon_\perp > 0$ and $\varepsilon_{//} < 0$, but for tensile strain are $\varepsilon_\perp < 0$ and $\varepsilon_{//} > 0$.

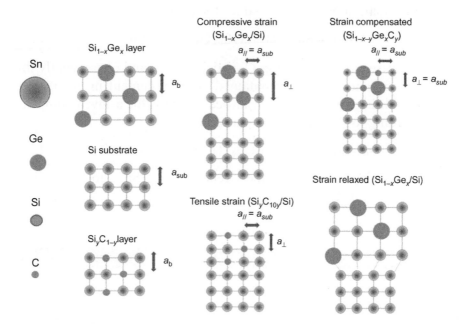

Figure 1.16 A schematic view of relaxed, strained, and strain-compensated group IV alloys.

In a similar way, mismatch parameters for crystalline layers are defined as follows:

$$f_\perp = \frac{a_\perp - a_b}{a_{sub}} \quad \text{and} \quad f_{//} = \frac{a_{//} - a_b}{a_{sub}} \qquad \text{(1.18a and 1.18b)}$$

The vertical mismatch for compressively strained crystals is $f_\perp > 0$ and $f_\perp < 0$ for tensile strain. For the strained materials, $f_{//} = 0$ when the layer is totally aligned to the substrate. The total mismatch can be obtained from $f_\perp$ and $f_{//}$ when the Poison ratio (ν) for the semiconductor is known [14–16].

$$f = (f_{//} - f_\perp)\frac{1-\nu}{1+\nu} + f_{//} \qquad (1.19)$$

where ν is a value which shows how a crystal responses elastically to exerted forces and is calculated for a certain layer from the elastic constants (Table 1.1) [21]:

$$\nu = \frac{c_{12}}{c_{12} + c_{11}} \qquad (1.20)$$

It is very important to determine the elastic constants for an alloy in order to calculate the Poison ratio. For instance, the following simple approximation provides C_{ij}-values for GeSn alloy:

$$C_{ij}(\text{GeSn}) = (1 - x)C_{ij}(\text{Ge}) + xC_{ij}(\text{Sn}) \qquad (1.21)$$

In this case, a corresponding C_{ij} is determined and consequently the Poisson ratio and f-value are obtained. The Sn content in the alloy is derived with high accuracy from lattice constant for a given composition according to the following equation [22]:

$$a_{\text{GeSn}}(x) = a_{\text{Sn}}x + \theta_{\text{SnGe}}x(1 - x) + a_{\text{Ge}}(1 - x) \qquad (1.22)$$

where θ_{SnGe} is a constant which relates to GeSn alloying and is 0.166 Å for $x \leq 0.20$.

Table 1.1 The Elastic Constants for the Group IV Materials				
Elastic Constant	Ge [17]	Sn [18]	Si [19]	C [20]
c_{11} (Mbar)	1.26	0.69	1.67	10.79
c_{12} (Mbar)	0.44	0.29	0.65	1.24

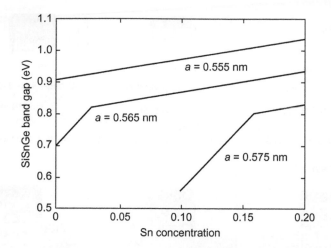

Figure 1.17 The bandgap change in ternary $Ge_{1-x-y}Si_xSn_y$ alloys with the same lattice constant [23].

The relaxation amount (R) is an important value to quantify the reduced strain upon relaxation. R is obtained according to $R = \frac{f_\parallel}{f}$. The R-value is expressed in percentage.

The (compressive and tensile) strain can be compensated if smaller/larger atoms are introduced in lattice to counterpart the induced compressive/tensile strain, respectively. Ternary alloys like $Si_{1-x-y}Ge_xC_y/Si$ and $Ge_{1-x-y}Sn_xSi_y/Ge$ demonstrate strain compensation.

The lattice constant of $Ge_{1-x-y}Si_xSn_y$ can be obtained from the following equation:

$$a_{Ge_{1-x-y}Si_xSn_y} = a_{Ge} + \Delta_{SiGe}x + \theta_{SiGe}x(1-x) + \Delta_{SnGe}y + \theta_{SnGe}y(1-y)$$

$$(1.23)$$

where $\Delta_{SiGe} = a_{Si} - a_{Ge}$ and $\Delta_{SnGe} = a_{Sn} - a_{Ge}$.

One of the excellent properties of the $Ge_{1-x-y}Si_xSn_y$ ternary system is the possibility to tailor separately the lattice constant and the bandgap. The bandgap may vary over a large range by tuning the Si/Sn ratio in the layers. A linear relationship between the bandgap and the Sn content is obtained for a certain lattice constant as shown in Figure 1.17. Lattice-matched $Ge_{1-x-y}Si_xSn_y$ ternary system can be grown strain-free for detection of a wide range of wavelengths [23].

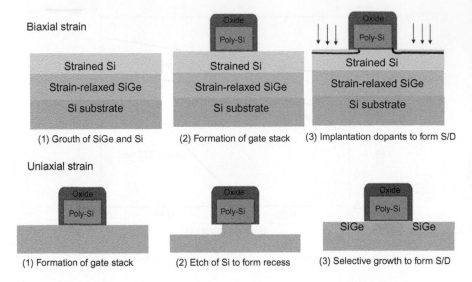

Figure 1.18 The process of formation of biaxial and uniaxial strained channel in a MOSFET structure.

STRAIN DESIGN FOR MOSFETs

There are two ways to induce strain depending on how the force is exerted: biaxial (global or 2D) and uniaxial (1D). Figure 1.18 illustrates how these types of strain are generated in silicon material.

An example for biaxial strain is when a tensile-strained Si layer is grown on strain-relaxed SiGe. A problem with such structures is that a graded SiGe layer (1 μm, 10% Ge) is used as a virtual substrate. This means that a few micrometers of SiGe are required to achieve high strain amount in the Si cap layer.

The purpose of inducing high strain amount in transistor structure is to decrease both the scattering and effective mass of carriers in the channel region.

In general, strain decreases the intraband scattering rate (where acoustic phonons are involved) by changing the density of states of LH and HH bands. The other effect of strain is to decrease the interband scattering (where optical phonons are involved) by splitting the HH and LH bands. This scattering rate is significantly reduced when the split of HH−LH bands is comparable to the energy of optical phonons. Therefore, in order to improve the carrier mobility, a stress amount larger than 1 GPa which corresponds to $\sim 25-30\%$ Ge is needed [24,25].

STRAIN EFFECT ON CARRIER MOBILITY

The carrier mobility is one of the most important topics for MOSFET design. The carrier mobility is affected by the presence of defects, interfacial states (e.g., at interface of high-k materials), surface roughness, dopant concentration, channel material, and channel strain. Some basic definitions are necessary to be provided here for further advanced discussions later in the chapter.

BASIC DEFINITIONS

The carrier movement under the influence of an electric field is described by the following equation:

$$\frac{d}{dt}(m^*v_d) + \frac{m^*v_d}{\langle\tau\rangle} = qE \tag{1.24}$$

where v_d is the drift velocity, E is the electric field, m^* is the effective mass, and τ is the scattering time for the carriers. In steady state, the first term of the above equation becomes zero. Then

$$\frac{m^*v_d}{\langle\tau\rangle} = qE \quad \text{or} \quad v_d = \mu E$$

where the mobility is defined as $\mu = q\langle\tau\rangle/m^*$.

In order to enhance the carrier mobility, either the effective mass has to be reduced or τ to be increased. The effective mass can be changed when the curvature of the subband is changed. This can be designed by inducing strain in the crystal. Meanwhile τ is changed when the scattering of carriers during the transport in the channel is decreased. The main scattering mechanisms in the semiconductors are addressed in the following.

Defect scattering: The carriers are scattered over the lattice deformations due to the lattice potential nonuniformities. There are two categories of defects: extended defects, e.g., dislocations and stacking faults, and local defects, e.g., point defects and precipitates (or islands) in the crystal lattice. The extended defects are mostly generated as a result of strain relaxation or problems during epitaxy process. The dopants or alloying materials also induce carrier scattering and therefore are often included in the defect scattering category.

Lattice scattering: The carrier scattering due to lattice deformations or phonon vibrations in acoustic or optical forms. The acoustic phonons have energies $\hbar\omega \ll kT$, where scattering rate is low, whereas optical phonons have energies $\hbar\omega > kT$ in vibrational dispersion relationships.

Deformation potential: When the acoustic phonons vibrate the lattice atoms, the conduction and valence band edges change as well.

Piezoelectric scattering: This occurs when a local electric field forms in the lattice once its potential changes by lattice vibrations in a way that more charge accumulates in one side of the lattice. Because of the polar nature of III−V compound semiconductors, piezoelectric scattering is dominant in these materials compared to the group IV materials. In MOSFETs the carrier mobility for a drive current is derived from expression 1.15 by inserting the oxide capacitance, physical dimensions of the gate and applied voltages to the transistor terminals.

CARRIER MOBILITY IN MOSFETs WITH STRAINED Si CHANNEL

The presence of strain changes the band structure in a way that both the carrier scattering and effective mass are reduced.

The early works on strain engineering in MOSFETs were focused on biaxially strained Si channel. The driving force behind this type of transistor design was to improve the carrier mobility for both p- and n-MOSFETs at the same time [26].

To induce global strain over a wafer, a series of developments were performed by wafer manufacturers which provided strained silicon and germanium on insulator (sSOI and sGOI). Another solution for advanced strain engineering is given by SiGe on insulator (SGOI). These layers are strain-free and can be used as virtual layers for highly strained Si or SiGe layers. Dual-channel heterostructures on insulator (DHOI) is one way to process nMOSFET and pMOSFET. Figure 1.19(a)−(d) illustrates such a SGOI wafer for DHOI design where the strained Si and SiGe layers are integrated to host the hole and electron channels, respectively, in high mobility devices (as described in Table 1.2) [26].

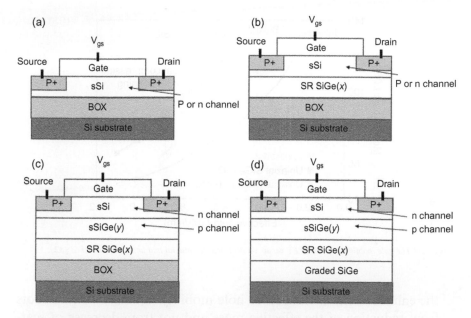

Figure 1.19 Different designs of biaxial-strained Si channel in MOSFETs [26].

Table 1.2 Descriptions of the Biaxial-Strained Si in MOSFETs	
Strained Si on insulator (sSOI) [27]	Commercially available substrates, high electron and hole mobility, no leakage between the transistors, low power consumption
SiGe on insulator (sGOI) [28]	Strain-relaxed SiGe is created through condensation technique, Si strained for high electron and hole mobility, no leakage between the transistors, low power consumption
Dual-channel heterostructure on insulator (DHOI) [29,30]	SiGe is created through condensation technique, the thicknesses can be reduced to minimize SCEs (advanced strain engineering)
Dual-channel heterostructure on bulk (DHOB) [31]	Strain-relaxed SiGe is created through epitaxy and annealing treatment high electron and hole mobility

Late in the 1990s, different analysis suggested that the industry process should be adapted to uniaxial strain. The main reasons are discussed as follows:

1. Uniaxial stress offers considerably larger hole mobility at both high and low applied electric fields because of the difference between the surface confinement for HL and HH bands (Figure 1.20).
2. Larger drive current can be obtained for uniaxial strain which is beneficial for short channel transistors. This is due to the fact that

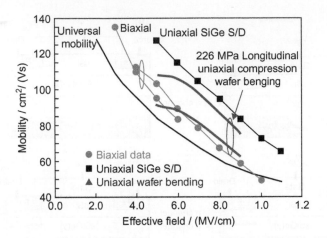

Figure 1.20 The hole mobility of pMOSFET versus vertical field for biaxial and uniaxial strained Si [32].

the enhancement in electron or hole mobility in uniaxial strain roots from reduction of the effective mass and not from decrease of scattering processes as is the case for biaxial strain.

3. The uniaxial stress provides almost five times smaller threshold voltage shift (ΔV_T) in n-channel MOSFETs. This is a very important point as this shift is usually tailored with the change of doping level in the channel region [32]. Equations (1.1a), (1.1b), and 1.2 show the relationship of threshold voltage with conduction band edge E_c, the density of states in valence band, N_V, and bandgap, E_g, as follows:

$$q\Delta V_T(\sigma) = \Delta E_c(\sigma) + (m-1)\left[\Delta E_g(\sigma) + kT\ln\frac{N_V(0)}{N_V(\sigma)}\right] \quad \text{Biaxial stress}$$

(1.25)

$$q\Delta V_T(\sigma) = (m-1)\left[\Delta E_g(\sigma) + kT\ln\frac{N_V(0)}{N_V(\sigma)}\right] \quad \text{Uniaxial stress} \quad (1.26)$$

According to the above equations, there are two reasons for the threshold voltage shift behavior:

1. Biaxial tensile stress results in a larger valence offset and larger bandgap narrowing in Si channel compared to uniaxial strain. As a result, LH band is maximum in biaxial stress, while HH is for uniaxial stress.
2. In uniaxial strain, the ΔE_c term makes up just a minor change since only the gate region is strained (Figure 1.21).

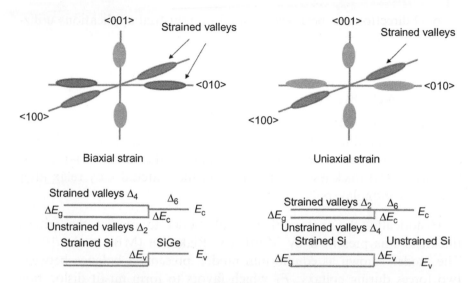

Figure 1.21 The band alignment of strained Si in biaxial and uniaxial strain [25].

In conclusion, these discussions indicate that uniaxial stress has more advantages for logic technologies compared to biaxial stress. Uniaxial strain was integrated for the first time at 90 nm node technology and will be the dominant strain type on the path of more Moore.

For industrial applications, three methods have been applied to increase the mobility: embedded or elevated SiGe-based stressors in S/D regions, SiN stress liners on the transistor and hybrid substrates. The hybrid substrate refers to wafers, e.g., with (110) orientation. High-performance ICs in many technological nodes has integrated these methods for nMOSFET and pMOSFET [33,34].

To explain the transport properties in the channel region, piezoresistance coefficients are indicators which are defined in terms of mobility's fractional variations as $\Delta\mu/\mu = \approx |\pi_{//}\sigma_{//} + \pi_{\perp}\sigma_{\perp}|$, where $\sigma_{\perp}$ and $\sigma_{//}$ are the transverse and longitudinal stresses, and $\pi_{//}$ and $\pi_{\perp}$ stand for the piezoresistance coefficients in longitudinal and transverse directions. The piezoresistance coefficients can be expressed in terms of fundamental cubic piezoresistance coefficients $\pi 11$, $\pi 12$, and $\pi 44$ or their combinations for a certain orientation [35].

The piezoresistance coefficient for holes is highest for compressive strain along $\langle 110 \rangle$ for both (001) and (110) wafers. Therefore, $\langle 110 \rangle$

channel direction has been widely used in industrial applications utilizing strained Si technologies [33,36].

STRAIN AND CRITICAL THICKNESS

Global Critical Thickness of SiGe Layers

All types of strained alloys store mechanical energy which may relax after their thickness exceeds a critical value. In the case of a large misfit, this critical thickness is very thin and the material may relax after only a few monolayers.

Historically, the first mathematical model to predict the critical thickness was presented by Matthews−Blakeslee (MB) in the 1970s. The model, known as equilibrium model, presents a balance between two forces during epitaxy: F_T which favors to form misfit dislocation and F_a to relief this force during epitaxy. If $F_a < F_T$, the strained material is deposited without any formation of dislocations. A critical thickness is calculated when $F_T = F_a$. The critical thickness is given by [37]:

$$h_c = \frac{1 - \nu \cos^2 \theta}{1 + \nu} \frac{b}{8\pi f \cos \lambda} \ln \frac{\alpha h_c}{b} \qquad (1.27)$$

A few years later when the epitaxial technique for SiGe layers was developed, a large discrepancy between the theoretical calculations and experimental data was observed. One way to explain this large discrepancy is that the MB theory takes into account only a balance between two acting forces to form misfit dislocations and it lacks consideration of the interaction, nucleation, and propagation of misfit dislocations. Therefore, the critical thickness for SiGe layers was remarkably underestimated.

The experimental values were later modeled and fitted by People and Bean [38]:

$$h_c = \frac{1 - \nu}{1 + \nu} \frac{\sqrt{2}}{32\pi} \frac{b^2}{af^2} \ln \frac{h_c}{b} \qquad (1.28)$$

A more sophisticated model was introduced by Dodson and Tsao (DT) when the concept of equilibrium forces was replaced by excess stress σ_{ex} between the mentioned forces in MB [39,40]. Therefore, the

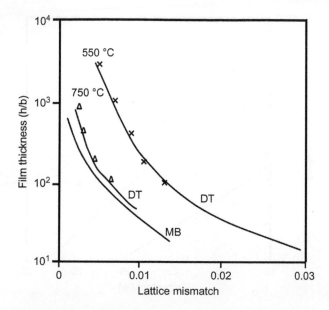

Figure 1.22 The calculated critical thickness of SiGe layers versus Ge content from DT model [39,40], where MB is from the equilibrium model [37]. The (×) and (Δ) marks are experimental data for the grown SiGe layers [38].

strain relaxation may occur if the nonzero σ_{ex} is set to zero. The **MB** equation is then retrieved again. σ_{ex} is expressed by

$$\frac{\sigma_{ex}}{\mu} = 2\varepsilon \frac{1-\nu}{1+\nu} - \frac{b}{2\pi h} \frac{1-\nu\cos^2\theta}{1-\nu} \ln\frac{4h}{b} \qquad (1.29)$$

where μ is the shear modulus and ε is the elastic strain. Figure 1.22 illustrates the experimental data reported from different groups and the calculated thickness for SiGe layers from DT and MB theories (when a critical $\sigma_{ex}/\mu = 0.026$ is reached). The region indicated between MB and DT thickness is called metastable. Most of the strained layers are grown with a thickness which places them in the metastable zone. A good consistency between the layer thickness and Ge content is observed. The metastable region for SiGe is a result of both equilibrium and nonequilibrium theories which considers the generated stress. In general, the stress in the epi-layers is temperature dependent. This means that the metastable region may become smaller or larger for high and low growth temperatures.

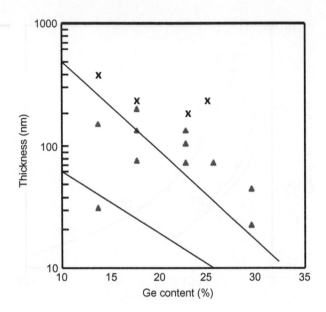

Figure 1.23 The experimental data showing thickness of strained (▲) and partially strain-relaxed (✕) SiGe versus Ge content [41].

CRITICAL THICKNESS OF SiGe LAYERS ON PATTERNED SUBSTRATES

The above experimental data and theories are for the SiGe layers grown on the whole wafer. The kinetics of the growth is quite different if the SiGe layers are grown selectively on the patterned substrates [41].

On a patterned substrate comprising openings in the oxide layer grown on top of a Si wafer, SiGe is strained in the center of oxide openings and is (partially) relaxed in the areas close to the oxide walls. This kind of strain distribution creates a media where the misfit dislocations may get depleted in the relaxed area. The presence of relaxed areas will also change entirely the kinetics of misfit dislocations and their propagation. This provides the possibility to grow strained SiGe layers selectively far above the critical thickness as shown in Figure 1.23. The presence and kinetics of defect propagation is affected in the presence of strain.

CRITICAL THICKNESS OF SiGe LAYERS GROWN ON NANO FEATURES

The recent reports have demonstrated that the strain relaxation of SiGe is delayed once grown on nanoscaled areas [42]. In FinFET

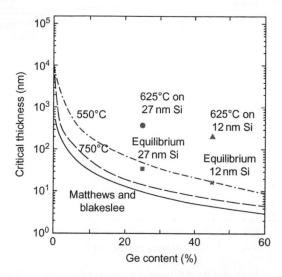

Figure 1.24 The critical thickness of SiGe layers grown on Si fins with different sizes [42].

structure, SiGe is grown on the Si-fin to raise the S/D areas. In this case, high-quality SiGe with a thickness far beyond the reported critical thickness for bulk material is observed. A reason for this behavior is that the strain forces on the fin are dependent both on the thickness of the Si fin (substrate) and the thickness of the grown layer. Since the strain relaxation of SiGe layers occurs through the formation of 60° dislocations which glide on {111} planes, the shape of the Si fin has an important role in the strain relaxation process. Therefore, the maximum-induced strain changes in terms of the fin/film thickness as well and is not directly related to the film thickness. As a result, the critical thickness of SiGe layers grown on Si fin is significantly increased compared to the bulk form as shown in Figure 1.24.

STRAIN MEASUREMENTS AND APPLICATIONS

Strain Measurement
There are three powerful techniques to measure strain in semiconductors: Raman spectroscopy, transmission electron microscopy (TEM), and high-resolution X-ray diffraction (HRXRD).

Raman Spectroscopy
This technique is a nondestructive method to measure strain in semiconductors. The technique is based on inelastic scattering of light

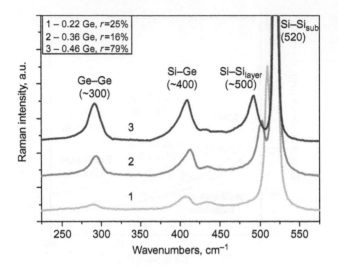

Figure 1.25 Raman spectra for strained SiGe/Si samples with different Ge contents.

(wave number in cm^{-1}) as a result of lattice vibrations and electronic excitations. The frequency shift of scattered light can be determined by comparing the signal from unstrained and strained layers. The frequency shift is interpreted to strain amount in the layer.

For a SiGe/Si structure, the three peak positions, Si–Si, Si–Ge, and Ge–Ge phonon modes, will be of interest as shown in Figure 1.25 [43].

This analysis offers a micro-Raman mapping which can measure the strain variations over an area with 1 μm resolution. The technique can be combined with AFM in order to obtain an accurate position on the sample [44].

In a Raman spectra, the relative energies and intensities of the Si–Si, Si–Ge, and Ge–Ge vibrations are dependent on the relative number and distortion of corresponding bonds in the alloy due to composition and strain, respectively [45,46]. This dependency of phonon frequency in terms of strain amount in SiGe layers, ε, is given by

$$\omega_{SiSi} = 520.2 - 62x - 815\varepsilon \tag{1.30}$$

$$\omega_{SiGe} = 400.5 + 14.2x - 575\varepsilon \tag{1.31}$$

$$\omega_{GeGe} = 282.5 + 16x - 385\varepsilon \tag{1.32}$$

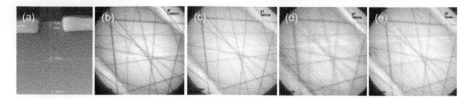

Figure 1.26 (a) Dark field image of a test structure with SiGe in S/D and CBED patterns at different distances from the top, (b) 1000 nm, (c) 500 nm, (d) 100 nm, and (e) 25 nm.

The third term (containing ε) is associated to strain where the first and second terms relate to relaxed SiGe material [44–47].

TEM Analysis

The convergent beam electron diffraction (CBED) in TEM is a technique to measure the local strain in nanoscale resolution. The technique has been applied to measure the strain in MOSFET's channel region. The main point is using high-order Laue zone (HOLZ) deficient lines which are sensitive to variations of lattice parameter. The strain is estimated when the Holtz patterns are simulated and compared to experimental results. Figure 1.26(a)–(e) illustrates how the HOLZ lines are changed when the strain is decreased on the path away from the channel. The accuracy of CBED technique is 2×10^{-4}.

A concern about applying CBED pattern method is the strain relaxation when e-beam is emitted to an area. The strain relaxation appears in the form of split of HOLZ lines and has to be taken into account in the simulations.

The other major problem with using CBED patterns is that the sample area has to be at least 200 nm thick in order to obtain a sharp pattern. This means that it is difficult to apply CBED for analysis of MOSFETs with ultrathin channel length. Therefore, in many cases, a test structure is analyzed and the results are correlated afterward to smaller dimensions [48].

High-Resolution X-Ray Analysis

Among the characterization techniques, XRD is the most popular one since it is a fast and a nondestructive method. X-ray rocking curves (RCs) are one-dimensional (1D) scans around a certain reflection in the reciprocal space. RCs are also known as $\omega - 2\theta$ scans, where ω and 2θ are incident and diffracted beams, respectively. The analysis is based on the Bragg law, $2d \sin \theta = n\lambda$, where d is the inter-planar distance, θ is the incident X-ray beam, and λ is the wavelength of $K_{\alpha 1}$ of copper which is 0.15406 nm. The strained layers experience a distortion in lattice constant in the z-direction, where d is either elongated or shrunk depending on the type of strain. When a multilayer structure contains layers with different strain amounts and types, θ shifts according to d-values and different peaks are generated during scanning. The position of layer peaks relative to the substrate peak reveals the nature and amount of strain of epi-layers. A series of fringe peaks may also appear in RCs which are generated from the interference of X-ray beam within the strained layers. The material composition is measured from layer and substrate peak positions and the layer thickness from two neighboring fringes in the RC.

The defect density can be estimated by measuring the full-width-half-maxium (FWHM) of peaks in the RC.

The misfit parameters ($f_\perp$ and $f_{//}$) can be measured by high-resolution reciprocal space mappings (HRRLMs) which are 2D measurements around a certain reflection in reciprocal space [14−16]. Four input parameters (ω and 2θ for layer and substrate) are obtained from an HRRLM and misfit parameters can be calculated. HRRLMs in (113) reflection direction are the most sensitive ones for measurement of strain relaxation and defects. This is due to the low grazing angle of X-ray beam (ω is 2.8°) which covers a large area of sample during the measurement.

HRRLMs and RC scans can be performed to measure strain in SiGe over chips on patterned substrates when the X-ray beam could be focused on a large array of the MOSFETs. Recently, (113) HRRLMs have been applied to measure strain over chips with 22 nm transistor nodes as shown in Figure 1.27(a) and (b) [48]. The figure shows the HRRLMs of intrinsic and B-doped SiGe layers grown in recess S/D of transistors. The B concentration can be estimated by comparing the intrinsic and doped SiGe layers' RCs and measuring the layer shift. This shift is originated from the strain compensation of B in SiGe matrix and is interpreted to the dopant concentration.

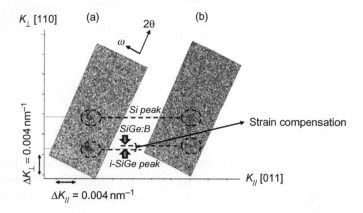

Figure 1.27 HRRLMS around (113) reflection of an array of 22 nm transistors containing $Si_{0.65}Ge_{0.35}$ layers in S/D region: (a) intrinsic layer and (b) B-doped with concentration of 3×10^{20} cm^{-3}. The shift of SiGe:B peak (toward Si peak) compared to i-SiGe is due to the strain compensation [49].

The drawback with X-ray technique is the impossibility of locally distinguishing the strain in the transistor structures; the technique can only provide information over a large array of devices.

PART THREE: CHEMICAL VAPOR DEPOSITION OF GROUP IV MATERIALS

CVD refers to a process in which certain precursor gases decompose over a substrate inside a reactant chamber at a specific growth temperature. There are several precursors for Si epitaxy depending on the growth temperature and application: silane (SiH_4), dichlorosilane (SiH_2Cl_2), trichlorosilane ($SiHCl_3$), tetrachlorosilane ($SiCl_4$), disilane (Si_2H_6), and trisilane (Si_3H_8). For Ge epitaxy, germane (GeH_4) and digermane (Ge_2H_6) are the common sources. There are also a variety of ($H_3Ge)_xSiH_{4-x}$ ($x = 1-4$) compounds for the growth of Ge-rich SiGe layers [50]. The dopant precursors are diborane (B_2H_6) for p-type doping and phosphine (PH_3) and arsine (AsH_3) for n-type doping. ($As(GeH_3)_3$ is a gas source which is also used for n-type doping of SiGe layers at low growth temperature [50]. These sources are usually diluted in hydrogen. For Sn, SnD_4 and $SnCl_4$ are the most practical and common sources. Methylsilane (SiH_3CH_3) is widely used for carbon doping in Si and SiGe layers.

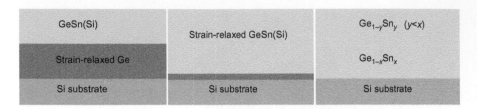

Figure 1.28 Schematic of structures where GeSn can be grown strained or strain-relaxed.

Ge−Sn−Si alloys are known for at least two decades for their potential for indirect-to direct bandgap transition [51]. At first, these alloys were epitaxially deposited by molecular beam epitaxy (MBE) [52]. However, the growth of these alloys using CVD technique was not demonstrated until recently when a stable Sn precursor, like SnD_4, was produced [53,54]. Although this precursor has demonstrated impressive results, the SnD_4 molecule has a limited lifetime and dissociates into Sn and deuterium after a while. The stability problem can be partly solved when SnD_4 is diluted in hydrogen prolonging the lifetime to a few months.

Another source for Sn is $SnCl_4$ which is widely used to grow Sn oxides by atomic layer deposition (ALD). This source is in liquid form, is converted into gas form, and later a stable gas flux of $SnCl_4$ molecules is introduced into the CVD reactor. GeSnSi layers with high Sn content and high layer quality have been deposited in temperature range of 290−380°C [53,54]. Figure 1.28 illustrates how to grow GeSn layers with different strain types.

SELECTIVE AND NONSELECTIVE EPITAXY

If a layer is deposited on the entire wafer, the epitaxy is called nonselective. Otherwise, a selective epitaxial growth (SEG) on patterned substrates can be performed where the growth takes place only on exposed Si areas inside oxide (or nitride) openings. This type of epitaxy occurs when HCl is introduced during deposition. The dissociated Cl atoms act as Si etchant and remove nucleated Si on the oxide (or nitride).

The SEG found a key application in MOSFETs and photonic components [55]. SiGe can be deposited selectively as stressor material in S/D regions to create uniaxial compressive strain in the transistor channel and enhance the hole mobility in pMOSFETs [32].

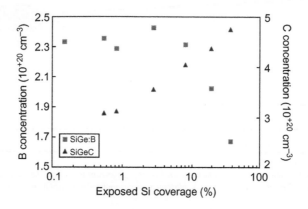

Figure 1.29 The variation of B and C concentration in SiGe layers grown selectively on chips with different coverage of exposed Si areas. Dichlorosilane, germane, methylsilane, HCl, and diborane partial pressures were 60, 1.2, 0.3, 20, and 3.6 mtorr, respectively [56].

Although selective epitaxy is a very attractive deposition technique for MOSFETs, this growth method suffers from facet formation, nonuniformity of layer profile inside the oxide openings (micro-loading), or over different parts of a chip or wafer (global loading or pattern dependency) [56–59].

The nonuniform deposition is also observed from wafer to wafer when the thickness of oxide or nitride on the wafers is changed. This growth problem leads to uncontrolled Ge content and growth rate (layer thickness) over the chip area. As a result, the incorporation of boron (or carbon) in SiGe is affected and the sheet resistance in S/D junctions will not be uniform as well (Figure 1.29).

Many investigations have been carried out to find methods to improve the uniformity of growth and to decrease the pattern dependency to <5% [57,58], but so far there is no remedy to totally eliminate this problem.

The reason for pattern dependency links to the gas kinetics and nonlinear consumption of the gas molecules over a patterned wafer [56,59]. The kinetics of CVD growth can be described by the established boundary stream theory assuming a laminar gas flow over the Si wafer as illustrated in schematic view in Figure 1.30 [59]. For the growth of SiGe layers in S/D regions in planar MOSFETs, the layout of the chip (size, shape, and density of openings) is an essential matter.

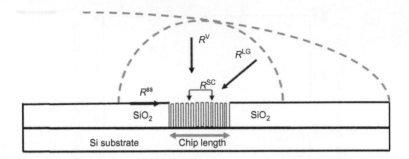

Figure 1.30 Schematic view of the gas boundary when laminar gas stream flows over the patterned substrate during the CVD process. The black arrows show the movement of gas molecules toward a chip [59].

The gas molecules move within streams and diffuse down through the boundaries. These diffused gas molecules are attracted toward the exposed Si areas and are consumed by dangling bonds. In the case of patterned substrates containing chips with opening arrays, a gas depletion volume is established as shown in Figure 1.30.

The total growth rate (R_{tot}) considering all the contributing components is expressed as

$$R_{Tot} = R_{Si}^V + R_{Si}^{LG} + R_{Si}^{SS} + R_{Si}^{SC} + R_{Ge}^V + R_{Ge}^{LG}$$
$$+ R_{Ge}^{SS} + R_{Ge}^{SC} - R_E^V - R_E^{LG} - R_E^{SS} - R_E^{SC} \qquad (1.33)$$

where R_V and R_{LG} represent the reactant gas molecules in vertical and lateral directions, while R_{SS} and R_{SC} denote the dissociated molecules on the oxide (or nitride) surface around or within a chip. The R_E terms stand for the etch rate caused by the HCl etchant species.

In order to reduce the pattern dependency of the growth, the lateral components in Eq. (1.33) have to be decreased.

Various methods have been suggested to reduce the pattern dependency of SiGe layers.

High HCl partial pressure decreases fast the density of species on the oxide surface, and the contribution of R_{SS} and R_{SC} components will be eliminated.

Another growth parameter which may affect the lateral components is the carrier gas velocity controlled by the total pressure in the reactor.

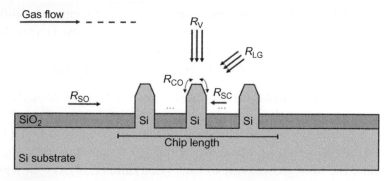

Figure 1.31 Schematic view of the gas boundary during the CVD process when the laminar gas stream flows over a patterned substrate containing Si-fin. The black arrows show the movement of gas molecules toward a chip [60].

When the thickness of gas depletion volume is decreased, a faster diffusion of the gas molecules to the surface is occurred. High partial pressure of H_2 carrier gas will also give a similar effect. Therefore, a combination of low growth pressure (10 torr) and high H_2 carrier gas (40 slm) will solve the pattern dependency behavior [59].

For the 3D growth of SiGe over Si-fins or Si wires (for 14 nm technology node and beyond) Eq. 1.33 is modified by adding a component for diffusive molecules over Si surface (R_{CO}) (Figure 1.31). The Si fins can be processed for different shapes, e.g. rectangular, trapezoid. In this case, such fins may contain a series of facet planes (for a trapezoidal fin has (100) and (111) planes) and during epitaxy new facet planes are formed at the sides and bottom side of the fin as well.

The total growth rate for 3D growth is written as [60]

$$R_{\text{Total}} = R_{\text{Si}}^{\text{V}} + R_{\text{Si}}^{\text{LG}} + R_{\text{Si}}^{\text{SO}} + R_{\text{Si}}^{\text{CO}} + R_{\text{Si}}^{\text{SS}} + R_{\text{Ge}}^{\text{V}} + R_{\text{Ge}}^{\text{LG}} + R_{\text{Ge}}^{\text{SO}}$$
$$+ R_{\text{Ge}}^{\text{CO}} + R_{\text{Ge}}^{\text{SS}} - R_{\text{HCl}}^{\text{V}} - R_{\text{HCl}}^{\text{LG}} - R_{\text{HCl}}^{\text{SO}} - R_{\text{HCl}}^{\text{CO}} + R_{\text{HCl}}^{\text{SS}} \tag{1.34}$$

According to the diffusion theory, the atoms on inclined planes have a longer diffusion length compared to (001) plane. Therefore, in 3D growth a flow of atoms from inclined planes at the sides is formed towards the (001) plane in central part of Si-fins [60].

However, a diffusion length of hundreds nanometers is estimated for Si and Ge atoms for growth at $600-700\,°C$, which is significantly larger than the width of a Si fin. This means the atoms may diffuse over the entire Si fin. Since the diffusion length of atoms is very large no interaction between the facet planes on the top and bottom of fin is expected. In this case R^{CO} term can be neglected in eq. (1.34).

However, if the dimensions (x, y and z) of the fin was enough large then the kinetics over the contribution of R^{CO} term in eq. 1.34 should be taken into account [60]. Calculating all the components in Eqs. (1.33) and (1.34) needs a tedious work. An empirical model has been applied to calculate the total growth rate in Eqs. (1.33) and (1.34) as following:

Free dangling bonds   Effect of SiH$_2$Cl$_2$

$$
\begin{aligned}
R_{Total} = \beta & \frac{(1 - \theta_{H(Si)} - \theta_{Cl(Si)})}{N_0} \frac{P_{SiH_2Cl_2}}{(2\pi m_{SiH_2Cl_2} k_b T)^{\frac{1}{2}}} \left(\frac{E_{SiH_2Cl_2 on \, Si}}{k_b T} + 1\right) \exp\left(-\frac{E_{SiH_2Cl_2 \, on \, Si}}{k_b T}\right) \\
+ \chi & \frac{(1 + m_r)(1 - \theta_{H(Si)} - \theta_{Cl(Si)})}{N_0} \frac{P_{GeH_4}}{(2\pi m_{GeH_4} k_b T)^{\frac{1}{2}}} \left(\frac{E_{GeH_4 on \, Si}}{k_b T} + 1\right) \exp\left(-\frac{E_{GeH_4 on \, Si}}{k_b T}\right) \\
+ \chi & \frac{(1 + m_r)(1 - \theta_{H(Si)} - \theta_{Cl(Si)})}{N_0} \frac{(BP_{GeH_4} \ln(1/c))}{(2\pi m_{GeH_4} k_b T)^{\frac{1}{2}}} \left(\frac{E_{GeH_4 on \, Si} + 0.1 \, eV}{k_b T} + 1\right) \exp\left(-\frac{E_{GeH_4 on \, Si} + 0.1 \, eV}{k_b T}\right) \\
- \frac{\gamma}{N_0} & \frac{P_{HCl}^{0.596}}{(2\pi m_{HCl} k_b T)^{a\frac{1}{2}}} \left(\frac{E_{Etching}}{k_b T} + 1\right) \exp\left(-\frac{E_{Etching}}{k_b T}\right)
\end{aligned}
$$

← Effect of GeH$_4$

Effect of HCl

$$(1.35)$$

where E and P stand for activation energy and partial pressure of the reactant gases, and θ parameters denote the occupied dangling bonds in presence of H and Cl on Si with a specific surface orientation and temperature. The constants β, χ, and γ are tooling factors which are dependent on gas kinetics and the temperature distribution in the CVD reactor. In Eq. (1.35), the term c relates to the pattern layout of the chip which is the main factor for the pattern dependency behavior of the growth.

During selective epitaxy, the Cl atoms remove the nucleated islands on the oxide (nitride surface). In this etch process, Cl reacts mostly with Si. As a result, Ge atoms are the most probable atoms on the oxide surface. To have this considered, an additional energy term of 0.1 eV has been added to germane activation energy to consider the diffusion of Ge on oxide surface. In Eq. (1.35), m_r is Ge reaction factor on Si surface and is estimated to be 2. This factor takes into account the influence of Ge atoms to enhance the number of the dangling bonds. Eq. (1.35) can be used for both 2D (to fill out the recess or elevate source/drain) and 3D (Si fins) growth. The difference between 2D

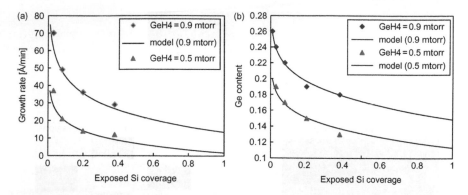

Figure 1.32 (a) Growth rate and (b) Ge content versus chip exposed Si coverage for different globally patterned wafers. The applied total pressure was 20 torr and partial pressures of HCl and germane were 60 and 20 mtorr, respectively [61].

and 3D growth is the input parameters. In 3D growth, the Si fin contains facet planes where the available dangling bonds has be to modified.

The Ge content (x term) in SiGe layers is acquired from a ratio of Si and Ge partial pressures. This is expressed as follows:

$$\frac{x^2}{1-x} = \alpha \exp\left(\frac{0.7 \text{ eV}}{k_b T}\right)\left(\frac{P_{\text{GeH}_4} + (BP_{\text{GeH}_4} \ln(1/c)) - (1-\lambda)P_{\text{HCl}}}{P_{\text{SiH}_2\text{Cl}_2} - \lambda P_{\text{HCl}}}\right)$$

$$(1.36)$$

where λ is a reaction ratio which represents the fraction of Cl atoms which interact with Si and Ge atoms. The value of this ratio is close to 1 for $P_{\text{HCl}} < PS_{\text{iH2Cl2}}$ and is 0.8 for high HCl partial pressures.

Equations (1.35) and (1.36) have been applied to calculate the SiGe profile grown on chips with different layouts. Figure 1.32(a) and 1.32 (b) shows a good consistency of the calculated and the experimental data.

All these calculations are for chips with uniform layout (according to Figure 1.32(a) and 1.32(b)), but this is an ideal case and for advanced chips, the layout is not uniform. A series of investigations have applied an interaction theory between neighboring chips on a Si wafer. The main equation comes from an empirical model which shows the variation of growth rate. This equation has an exponential behavior in terms of the distance, d, from a chip where the whole

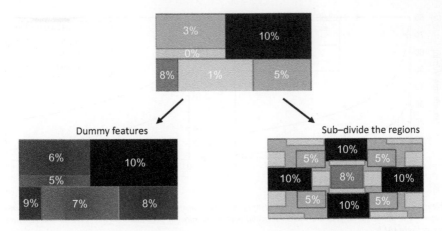

Figure 1.33 Modification of pattern distribution to obtain a uniform SiGe deposition [61].

interaction range is estimated to be τ. In this situation, the gas consumption becomes nonuniform over a chip or from chip to chip. On a patterned substrate, the dominant chips (R_{trap}) are the ones with more exposed Si area compared to surrounding chips (R_{Surr}). The growth rate of a chip under influence of a trap chip is $R_T(d)$ [61]:

$$R_T(d) = R_{Trap} + (R_{Surr} - R_{Trap})\left(1 - e^{\frac{-d}{\tau(c_{surr})}}\right) \qquad (1.37)$$

The interaction length is dependent on the coverage of the chip since they are the dangling bonds which attract the gas molecules toward themselves. In order to have a uniform growth over a chip, uniform gas consumption is required. Figure 1.33 illustrates a nonuniform chip and the way to change the layout to achieve uniform deposition. Six regions are distinguished in the figure with exposed Si coverage areas of 0%, 1%, 3%, 8%, and 10%. The interaction between these regions has to be calculated mutually. The main trap regions are 8 and 10 in this chip. Uniform gas consumption over this chip is obtained either by introducing dummy features or by subdividing the trap regions over the chip. Another way to achieve the goal is to apply both methods.

PART FOUR: IMPROVEMENT OF THE CHANNEL MOBILITY

EFFECT OF RECESS SHAPE IN S/D

In MOSFETs, the channel mobility is directly linked to the strain amount in the channel region. Many investigations have paid attention on how

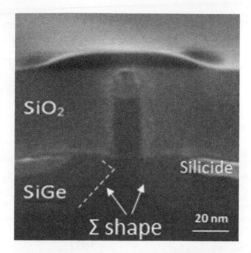

Figure 1.34 Sigma (Σ) shape recess in S/D of 22 nm bulk SiGe MOSFETs [49].

stress is exerted by SiGe in S/D region to the channel region. One way to increase the stress is to increase the Ge content. The Ge content has continuously been increased from 18% to 40% from 90 to 32 nm technology node. However, there is no more possibility for further increasing the Ge content since the defect density increases significantly due to strain relaxation in SiGe layers. The other way is to engineer the shape of the recess in S/D region. The idea behind is that the critical thickness and defect density depend on the surface facets in the recess bottom shape. This is explained by the difference of nucleation rate of SiGe layers on different facet planes. As an example, the SiGe layers deposited on the (110) facet plane show more roughness and have higher defect density compared to the (001) surface due to higher nucleation on (110) than (001).

The traditional U or square shape recess which has been used in some technological nodes consists of (110) surface and therefore is not optimized for highly strained SiGe layers [62].

In the later years, the recess shape was optimized to a sigma (Σ) shape S/D. This shape consists of (111) and (001) surfaces and was shown the most proper shape for pMOSFET application [48] (Figure 1.34).

The sigma shape increases the channel stress since the force to surface channel is from the sidewall toward the top surface. Figure 1.35 shows the effect of different recess shapes on the stress amount and carrier mobility [63].

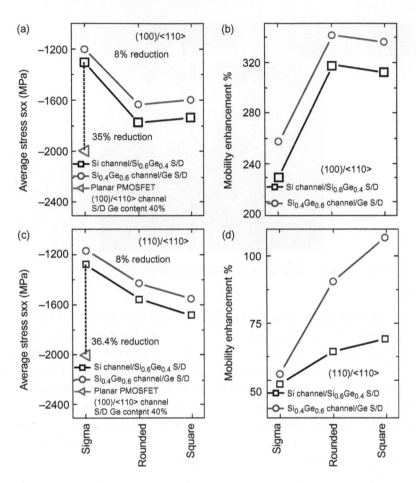

Figure 1.35 The stress and the corresponding mobility enhancement for different recess shapes for S/D region [63].

CHANNEL MATERIALS AND MOBILITY

III–V Materials

The carrier mobility in channel is an important characteristic of MOSFETs. As discussed in Chapter 1, strain can be induced in different ways in the channel region and improve the carrier transport, but it seems that the mobility enhancement cannot be continued for beyond 14 nm node. Thus, search for new channel material is ongoing to replace Si. Graphene and graphene-like materials with excellent mobility data could be good candidates for future transistors but at the moment being, they lack a reliable manufacturing method over the whole Si wafer.

Table 1.3 Material Data for Well-Known III–V Materials and Compares Them with Si and Ge						
Material	Si	Ge	GaAs	InP	InAs	InSb
Lattice constant (Å)	5.431	5.658	5.653	5.869	6.058	6.749
Bandgap (eV)	1.12	0.66	1.42	1.34	0.35	0.17
Hole mobility (cm^2/V-s)	450	1900	400	200	500	850
Electron mobility (cm^2/V-s)	1400	3900	8500	5400	40.000	77000

Some studies propose III–V materials (GaAs, InP, InAs, and InSb) as channel material [64]. The main idea behind is to use the low electron mass of III–V semiconductor due to their small bandgap (Table 1.3). The low electron mass property can be applied in high electron mobility transistors (HEMTs). These materials have a low density of states (DOS) resulting a low inversion charge and drive current. Due to large lattice mismatch between III–V materials and Si, the growth of such materials on Si with high epitaxial quality is difficult. It has been reported that a GeSnSi intermediate layer will solve this problem. The lattice constant of GeSnSi can be tailored by changing the Si and Sn content in Ge and is lattice-matched to a certain III–V material. The other issues with III–V integration on Si are their high dielectric constant and inefficient surface passivation. The first issue promotes SCEs in transistors and the latter results in high gate leakage. All these points indicate that using III–V materials as channel material is challenging.

Another group of material for HEMT is AlGaN/GaN structures [65–67]. These heterostructures generate a large 2D electron gas sheet density in the range of 10^{13} cm^{-2} without doping. Some research reports have proposed GaN-based high-power amplifier transistors for future 4G telecommunication application.

AlGaN/GaN structures are grown on Si when a nucleation layer in the range of micron is grown first as a buffer layer as shown in Figure 1.36. The lattice constant of AlGaN is less than GaN and thus in a AlGaN/GaN structures, a tensile near the interface of heterojunction is created. The tensile stress generates a static charge and a polarization field is formed due to piezoelectric properties of nitrides [65]. The metal gate forms a Schottky contact on GaN (AlGaN) surface.

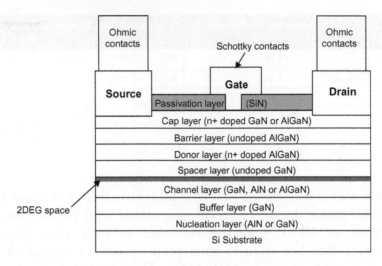

Figure 1.36 Layer structure of an AlGaN/GaN HEMT grown on Si [65–67].

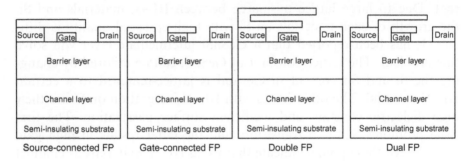

Figure 1.37 Different alternatives of field plates for GaN HEMTs [68].

Different field-plate designs have been proposed to improve the performance of the device. These methods make several features of the metallic electrodes extended over the gate and the gate–drain region [67], In this way, an electric field is established at the edge of the gate electrode from the drain side which leads to higher drain voltages and output power. A few alternatives of these methods are illustrated in Figure 1.37 [68].

Graphene Material

Graphene is a 2D crystalline structure in which carbon atoms are situated in a regular sp^2-bonded hexagonal pattern. Graphene is the building block of graphite and was for the first time isolated out of graphite in 2004 [69]. The planar structure and the ultimate thinness of graphene give rise to outstanding mechanical, electrical, and optical properties which can be utilized in future electronic—photonic devices. These properties and some of the implementations, so far, are discussed in the coming sections.

Graphene is produced in different ways. As mentioned above, it was first extracted from graphite by a surprisingly simple but groundbreaking method called mechanical exfoliation. In this method, the layers of multilayered graphite are split step by step until a single atomic layer is achieved. Geim and Novoselov used adhesive tape to do so and therefore this method is also often called "scotch tape method." The achieved single-layer slices are then deposited on a substrate [69,70].

Epitaxy deposition can also be used to produce graphene. Epitaxy is the process of making a crystalline layer on top of a crystalline substrate. As the outstanding qualities of graphene are best attained in isolation, a substrate with minimum interaction with graphene is usually sought. Silicon carbide (SiC), metals like iridium (Ir) or hexagonal boron nitride (hBN) have been used as substrates for graphene epitaxy.

Graphene is also obtainable by low-pressure heating of SiC wafers which results in epitaxial graphene. Reduction of graphite oxide, metal—carbon melt process, slicing of carbon nanotubes, and solvent exfoliation are among the other methods to produce graphene.

Due to its 2D feature, graphene has unique electronic properties. Intrinsic and large area graphene is a gapless semiconductor or a semimetal. Figure 1.38 shows that the dispersion relation is linear (conical in 3D) for low energies near the six corners of the 2D hexagonal Brillouin zone (Dirac points). Near these six points, both electrons and

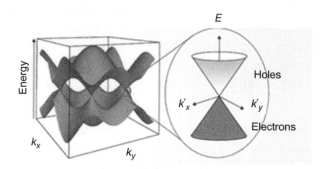

Figure 1.38 Electron energy band structure for a single-sheet graphene [71].

holes (Dirac fermions) have zero effective mass. They behave like relativistic particles described by the Dirac equation for spin-1/2 particles. The linear dispersion relation is given by

$$E = \hbar v_F \sqrt{k_x^2 + k_y^2}$$

where v_F is the Fermi velocity, k_x and k_y are wave vectors measured from the Dirac points. The Dirac points are considered as having the zero level of energy (E).

As expected from the low charge carrier effective mass in graphene [72], transport calculations and measurements show extremely high values for electron and hole mobility: 40,000 cm^2/Vs in graphene on SiO_2 [73], 27,000 cm^2/Vs in epitaxial graphene [74], and 200,000 cm^2/Vs in suspended graphene [75]. The mobility in the latest case is limited by intrinsic acoustic phonons of graphene and is higher than that of silver in room temperature. The choice and quality of substrate for unsuspended graphene can have a great effect on graphene's mobility as scattering by phonons of the substrate is dominant comparing to the graphene intrinsic phonons. Graphene is also capable of carrying high current densities [76] and its doping is possible by donor [77] or acceptor [78] atoms. Unlike conventional metals, the density of states and Fermi energy of graphene can be tuned by electrostatic or chemical doping [79].

Graphene has interesting potencies in three main fields of application involving its electronics properties: Biosensing, radio frequency switching, and in digital logic transistors. For the first two fields, numerous successful implementations are reported, although optimization and improvement have still room to go on. The main principle of sensing by graphene employs its high sensitivity to surrounding species. Many works are based on a structure often called "graphene transistor" in which, in principle, electronic current flows in gated graphene which in turn is in contact with species to be sensed. Upon interaction, electronic current (or conductivity, noise, etc.) in the device is influenced in a way that can be characteristic of particular species (Figure 1.39).

Graphene has also been reported to be successfully used as channel in FET for radio frequency switching with outstanding cutoff frequencies. The ballistic regime in graphene which takes place at room temperature, the high intrinsic carrier mobility, and tunable input

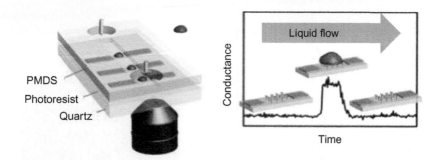

Figure 1.39 Graphene-based detection of single malaria-infected red blood cells by the changes they induce in conductivity of graphene [80].

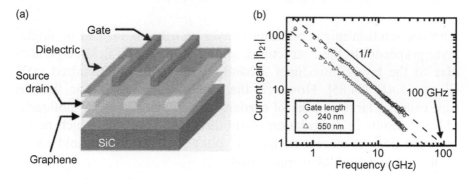

Figure 1.40 (a) Scheme of a microwave FET using graphene as its channel and (b) its current amplification gain by frequency. The cutoff frequency is ~50 and ~100 GHz for 550 and 240 nm long gates [82].

impedances around 50 Ω for graphene-based devices make it an interesting material choice for such devices [81] (Figure 1.40).

For digital logic MOSFETs, graphene's high mobility is tempting for getting very fast devices while its very small thickness can help avoiding SCEs like DIBL, threshold voltage roll-off, and impaired saturation of drain currents. It also offers higher electron velocities at high fields compared to Si or most III−V materials [83]. However, absence of a bandgap in graphene is a major obstacle that prevents simple off-switching of MOSFETs with graphene as channel material resulting in rather I_{on}/I_{off} ratios insufficient for digital logic. Therefore, more work around is required to introduce a bandgap to the design. One of the solutions is slicing the graphene into ribbons (graphene nanoribbons or GNRs). This can bring about a bandgap, but in price of dramatic decline in electron mobility. In addition, GNR processing to make perfect layers with high-quality edges is not straightforward.

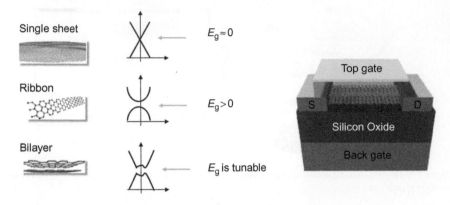

Figure 1.41 A bilayer graphene schematic and its bandgap diagram [84,87].

Another solution might be using bilayer or trilayers instead of single-layer graphene. In these structures, applying an electric field perpendicular to the layers introduces a bandgap of up to few hundred meV (Figure 1.41) [84,85]. However, the applied external electric field in turn causes a large number of carriers in the device, even in subthreshold and results in degradation of subthreshold slope [86] which represents the slope of current decline by voltage below the threshold and is important in digital logic applications.

Some other approaches to build a bandgap in graphene channeled FETs include graphene substrate interaction, lateral confinement, and straining [84], as well as introducing defects by ion irradiation [88]. Furthermore, alternative solutions by modifying the FET design (for instance by devising a physical gap along the channel [89]), totally new device concepts (like a graphene-based hot electron transistor [90]), and highly compatible gate dielectrics for graphene [91,92] have been proposed to secure high on−off current ratios with/without opening an electronic bandgap [93,94].

Silicene, Germanene, and Other Similar 2D Materials

Similar crystal structures to graphene can be thought of for other group IV elements and certain other materials. In analogy with graphene, single layers of atoms of silicon and germanium are named silicene and germanene, respectively, and are synthetically produced [95,96].

Silicene is not completely planar, as graphene is. It is buckled (Figure 1.42), meaning that some of the Si atoms are out of the plane.

Figure 1.42 Schema of Si atoms in a hexagon of buckled silicene (left) and a typical silicene cluster with ripples on the surface (right) [97].

This formation is due to larger atomic radius of Si atom and might be utilized as it produces an electronic bandgap which graphene lacks [98]. Moreover, silicene is more compatible with silicon device technology. Dirac fermions are also predicted for silicene [99], but their existence in silicene which is deposited on metallic substrate is debated [100,101]. It is believed that the mixing of orbitals of Si atoms with metal atoms on the substrate hinder their presence. Suspended or on-insulator silicene are expected to experimentally show the Dirac fermions, but they are yet to come.

Optical properties of silicene (and germanene) are predicted to be similar to graphene with the same low-frequency absorbance in IR range [102]. Hydrogenated silicene and germanene (known as silicane or germanane, respectively) are predicted to be wide bandgap semiconductors; germanane with a direct bandgap and silicane direct or indirect, depending on its atomic configuration [103].

BN, SiC, GaN, ZnO, MnO_2 MOFs, and transition metal dichalcogenides are some other monolayer materials. "The transition metal dichalcogenides are a class of materials with the formula MX_2, where M is a transition metal element from group IV (Ti, Zr, Hf, and so on), group V (for instance, V, Nb, or Ta) or group VI (Mo, W, and so on), and X is a chalcogen (S, Se, or Te)" [104]. They possess a wide range of physical properties. In contrast to graphene, they have a sizeable electronic bandgap which can be a crucial advantage. Molybdenum disulfide (MoS_2) is one such material which is naturally available in bulk form and can be exfoliated down to monolayers. In MoS_2, similar to other transition metal dichalcogenides, each metallic (Mo) atomic layer is sandwiched between chalcogen atomic layers (sulfur). Calculations show that MoS_2 is an indirect semiconductor with a bandgap of 1.2 eV between the top of valence band (Γ point) and the bottom of conduction band (between Γ and K points). The optical direct bandgap is situated at K point. In the monolayer, MoS_2 turns into a direct bandgap semiconductor with a gap of 1.9 eV, while the optical direct gap (at the K point) stays almost unchanged [104].

Germanium Material

Germanium is an attractive channel material for pMOSFETs due to its 4-times higher hole mobility compared to Si. The small bandgap of Ge makes higher intrinsic carrier concentration compared to Si [105] resulting in considerably higher diffusion current in Ge pn junctions. Moreover, the dielectric constant of Ge is higher than Si which makes it vulnerable to SCE and DIBL. Therefore, a transistor with Ge channel may demonstrate dominantly trap-assisted and band-to-band tunneling effects (TAT and BTBT, respectively) at high electric fields [105,106].

As examples, Ge(110) pMOSFET has shown a mobility value of 230 cm^2/Vs [107,108] (with $IrO_2-LaAlO_3$ gate dielectric) while the GOI transistors demonstrated a slightly higher mobility of ~ 300 cm^2/Vs [109]. These values are, respectively, a factor of 3 and 1.5 times higher compared to universal hole mobility for Si MOSFETs on Si (110) and SOI substrates.

The carrier velocity in Ge saturates at lower electric fields in comparison with Si, therefore the Ge material is probably beneficial for the high-field operation conditions.

Inducing uniaxial strain ($\sim$2GPa) in the channel region improves the drive current by 81%, which is 1.4 times higher compared with Si MOSFET with the same strain amount in the channel region [110].

One of the most important issues to obtain high hole mobility in germanium is the amount of defect density in the grown layers. An anisotropic surface reconstruction of Ge on Si(110) results in twins in the epitaxial layers [111].

Recently, it has been reported that GeSn alloys may have high quality and smooth layers on Ge(110) surface. These alloys have also demonstrated higher hole mobility compared to Ge [112,113]. Figure 1.43 shows a summary of the performance of Ge and GeSn alloys as channel material for pMOSFETs.

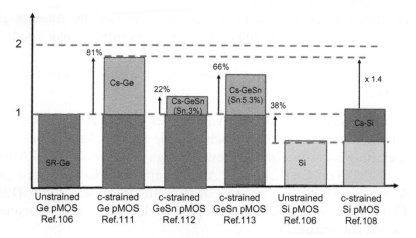

Figure 1.43 Impact of Ge and GeSn channel material (compared to Si) on the saturation drive current of a pMOSFET. The channel is strain-relaxed (SR) or compressive-strained (c-strained).

Due to large lattice constant difference between Ge and Si (4%) a high density of misfit and threading dislocations (TDs) can be generated. Different reports have demonstrated that the defect density can be reduced by post annealing treatment. For example, Lee et al has demonstrated that the TD density in a 1.5 μm thick Ge layer can be decreased from 10^8 to 10^7 after 3 min annealing at 800°C in nitrogen ambient [114].

Recent reports have demonstrated that good quality GeO_2 layers can be formed for passivation of Ge surface. The most important parameter for the quality of passivation oxide is density of interface traps (D_{it}) [traps/cm^2.eV].

A series of different oxidation methods e.g. conventional thermal oxidation [115], high-pressure oxidation [116−117], radical oxidation [118] and ozone exposure [119] have been applied to grow GeO_2. In these processes, it is important to prevent GeO_2 layer from direct contact with water moisture present in air. Therefore, the layers are capped by high-k materials [120] or rare earth oxides, e.g. Y_2O_3 [116−117].

The most common technique to deposit high-k material is atomic layer deposition (ALD). H_2O-free ALD process applying O_3 as an oxidant has been used to synthesize $HfCl_4/O_3$. The bandgap of GeO_2 bulk (H_2O assisted) is estimated to 4.3 eV and it was improved to values such as 5.4-5.9 eV for O_3-assisted grown layers.

The interface quality of GeO_2 can be improved by forming gas annealing (FGA) at 320–400 C. Higher temperatures may result in decomposition of GeO_2 layer through interaction with Ge surface. So far, D_{it} values in the range of $\sim10^{11}$ cm^{-2} eV^{-1} have been reported for GeO_2 [115–117].

Further research for suitable gate oxides for germanium has been also done. Fluorine- or sulfur-passivated Ge surfaces have been shown as interesting ways to directly grow the high-k materials (Ge/S/Al_2O_3/ HfO_2 or ZrO_2) on Ge [121–122]. Another approach for advanced gate stack for Ge MOSFETs is using rare earth oxides e.g. CeO_2 [123], La_2O_3 [124,125] or $LaLuO_3$ [116]. These oxides offer stable germanate layers and reasonable D_{it}.

Germanium is a suitable material for tunneling FETs (TFETs) due to the small bandgap. The carrier transport in TFETs occurs through tunneling of a source barrier and not diffusion over the barrier. High drive currents have been measured from Ge-TFETs which is comparable to conventional MOSFET [126].

In conclusion, there are different challenges in development of Ge-based CMOS but the on-going research activities may provide the possibilities to manufacture transistors with high performance for future industrial applications.

REFERENCES

[1] I. Ferain, C.A. Colinge, J.-P. Colinge, Multigate transistors as the future of classical metal−oxide−semiconductor field-effect transistors, Nature 479 (2011) 310–316.

[2] Hall, et al., Review and perspective of high-k dielectrics on silicon, J. Telecommun. Inf. Technol. 2 (2007) 33–43.

[3] E.S. Yang, Microelectronic Devices, McGraw-Hill Intern. Editions, ISBN 0-07-100374-6.

[4] D. Hisamoto, W.-C. Lee, J. Kedzierski, E. Anderson, H. Takeuchi, K. Asano, et al., A folded-channel MOSFET for deep-sub-tenth micron era, in: IEEE International Electron Devices Meeting Technical Digest, 1998, pp. 1032–1034.

[5] S.E. Thompson, In search of forever continued transistor scaling one new material at a time, IEEE Trans. Semicond. Manuf. 18 (2005) 26.

[6] M. Jurczak, N. Collaert, A. Veloso, T. Hoffmann, S. Biesemans, Review of FINFET technology, in: 2009 IEEE International SOI Conference, Foster City, California, 2009, p. 4.

[7] G. Pei, J. Kedzierski, P. Oldiges, M. Ieong, E.C. Kan, FinFET design considerations based on 3-D simulation and analytical modeling, IEEE Trans. Electron Devices 49 (2002) 1411–1419.

[8] A. Bachtold, P. Hadley, T. Nakanishi, C. Dekker, Logic circuits with carbon nanotube transistors, Science 294 (2001) 1317–1320.

[9] J. Appenzeller, J. Knoch, V. Derycke, et al., Field-modulated carrier transport in carbon nanotube transistors, Phys. Rev. Lett. 89 (2002) 126801.

[10] A. Dixit, et al., Analysis of the parasitic S/D resistance in multiple-gate FETs, IEEE Trans. Electron Devices 52(6) (2005) 1132–1140.

[11] T. Linton, M. Chandhok, B.J. Rice, G. Schrom, Termination of the line edge roughness specification for 34 nm devices, Tech. Dig. IEDM (2002) 303.

[12] T. Yamaguchi, H. Namatsu, K. Yamazaki, K. Kurihara, Nanometer-scale linewidth fluctuations caused by polymer aggregates in resist films, Appl. Phys. Lett. 71 (1997) 2388.

[13] J. Hallstedt, P.-E. Hellstrom, H.H. Radamson, Sidewall transfer lithography for reliable fabrication of nanowires and deca-nanometer MOSFETs, Thin Solid Films 517 (2008) 117.

[14] H.H. Radamson, J. Hållstedt, Application of high-resolution x-ray diffraction for detecting defects in SiGe(C) materials, J. Phys. Condens. Matter 17 (2005) S231517.

[15] G.V. Hansson, H.H. Radamsson, W.-X. Ni, Strain and relaxation in Si-MBE structures studied by reciprocal space mapping using high resolution X-ray diffraction, J. Mater. Sci. Mater. Electron. 6 (1995) 292.

[16] P.F. Fewster, X-ray Scattering from Semiconductors, Imperial College Press, London, 2000, ISBN: 1-86094-360-8.

[17] S.P. Nikanorov, B.K. Kardashev, Elasticity and Dislocation Inelasicity of Crystals, "Nauka" Publ. House, Moscow, 1985.

[18] P. Moontragoon, Z. Ikonić, P. Harrison, Band structure calculations of Si−Ge−Sn alloys: achieving direct band gap materials, Semicond. Sci. Technol. 22 (2007) 742–748. Available from: http://dx.doi.org/10.1088/0268-1242/22/7/012.

[19] S.P. Nikanorov, Y.A. Burenkov, A.V. Stepanov, Elastic properties of silicon, Sov. Phys. Solid State 13(10) (1971) 2516–2519.

[20] H.J. McSkimin, P. Andreatch, Elastic moduli of diamond as a function of pressure and temperature, J. Appl. Phys. 43 (1972) 2944–2948.

[21] J.J. Wortman, R.A. Evans, Young's modulus, shear modulus and Poisson's ratio in silicon and germanium, J. Appl. Phys. 36 (1965) 153–156.

[22] P. Aella, C. Cook, J. Tolle, S. Zollner, A.V.G. Chizmeshya, J. Kouvetakis, Structural and optical properties of $SnxSiyGe1-x-y$ alloys, Appl. Phys. Lett. 84 (2004) 888.

[23] J. Kouvetakis, J. Tolle, R. Roucka, V.R. D'Costa, Y.-Y. Fang, A.V.G. Chizmeshya, et al., Nanosynthesis of Si−Ge−Sn semiconductors and devices via purpose-built hydride compounds, ECS Trans. 16 (2008) 807–821. Available from: http://dx.doi.org/10.1149/1.2986840.

[24] C.W. Leitz, et al., Hole mobility enhancements and alloy scattering-limited mobility in tensile strained Si/SiGe surface channel metal−oxide−semiconductor field-effect transistors, J. Appl. Phys. 92 (2002) 3745–3751.

[25] M. Chu, Y. Sun, U. Aghoram, S.E. Thompson, Strain: a solution for higher carrier mobility in nanoscale MOSFETs, Annu. Rev. Mater. Res. 39 (2009) 203–229.

[26] A. Chaudry, G. Joshi, J.N. Roy, D.N. Singh, Review of current strained silicon nanoscaled MOSFET structures, Acta Tech. Napocensis Electron. Telecommun. 51 (2010) 15–22.

[27] S.H. Olsen, A.G. O'Neill, L.S. Driscoll, K.S.K. Kwa, S. Chattopadhyay, A.M. Waite, et al., High-performance nMOSFETs using a novel strained Si/SiGe CMOS architecture, IEEE Trans. Electron Devices 50 (2003) 1961–1969.

[28] D.J. Norris, A.G. Cullis, D.J. Paul, D.J. Robbins, High-performance nMOSFETs using a novel strained Si/SiGe CMOS architecture, IEEE Trans. Electron Devices 50 (2003) 1961–1969.

[29] S. Takagi, Understanding and engineering of carrier transport in advanced MOS channels, IEEE 52(2) (2008) 263–267.

[30] T. Mizuno, S. Takagi, N. Sugiyama, H. Satake, A. Kurobe, A. Toriumi, Electron and hole mobility enhancement in strained-Si MOSFETs on SiGe-on insulator substrates fabricated by SIMOX technology, IEEE Electron Device Lett. 21 (2000) 230–232.

[31] M.L. Lee, E.A. Fitzgerald, Optimized strained Si/strained Ge dual channel heterostructures for high mobility p- and n-MOSFETs, IEDM (2003) 18.1–4.

[32] S.E. Thompson, G. Sun, K. Wu, J. Lim, T. Nishida, Key differences for process-induced uniaxial vs. substrate-induced biaxial stressed Si and Ge channel MOSFETs, in: IEDM Tech. Dig., 2004, pp. 221–224.

[33] T. Ghani, S.E. Thompson, M. Bohr, et al., A 90 nm high volume manufacturing logic technology featuring novel 45 nm gate length strained silicon CMOS transistors, in: IEDM Tech. Dig., San Francisco, CA, 2003, pp. 11.6.1–11.6.3.

[34] S. Pidin, T. Mori, K. Inoue, S. Fukuta, N. Itoh, E. Mutoh, et al., A novel strain enhanced CMOS architecture using selectively deposited high tensile and high compressive silicon nitride films, in: IEDM Tech. Dig., Tokyo, Japan, 2004, pp. 213–216.

[35] C.S. Smith, Piezoresistance effect in germanium and silicon, Phys. Rev. 94 (1954) 42–49.

[36] M.D. Giles, M. Armstrong, C. Auth, S.M. Cea, T. Ghani, T. Hoffman, et al., Understanding stress enhanced performance in Intel 90 nm technology, in: VLSI Symp. Tech. Dig., Honolulu, HI, 2004, pp. 118–119.

[37] J.W. Matthews, A.E. Blakeslee, Defects in epitaxial multilayers. I. Misfit dislocations, J. Cryst. Growth 27 (1974) 118.

[38] R. People, J.C. Bean, Calculation of critical layer thickness versus lattice mismatch for Ge_xSi_{1-x}/Si strained-layer heterostructures, Appl. Phys. Lett. 47, 1985, p. 229.

[39] B.W. Dodson, J.Y. Tsao, Stress dependence of dislocation glide activation energy in single-crystal silicon-germanium alloys up to 2.6 GPa, Phys. Rev. B 38 (1988) 12383.

[40] B.W. Dodson, J.Y. Tsao, Scaling relations for strained-layer relaxation, Appl. Phys. Lett. 55 (1989) 1345.

[41] H.H. Radamson, A. Bentzen, C. Menon, G. Landgren, Observed critical thickness in selectively and non-selectively grown $Si1-xGex$ layers on patterned substrates, Physica Scripta, T101 (2002) 42.

[42] L. Yue, W.D. Nix, P.B. Griffin, J.D. Plummer, Critical thickness enhancement of epitaxial SiGe films grown on small structures, J. Appl. Phys. 97 (2005) 43519.

[43] T.S. Perova, J. Wasyluk, K. Lyutovich, E. Kasper, M. Oehme, K. Rode, et al., Composition and strain in thin $Si_{1-x}Ge_x$ virtual substrates measured by micro-Raman spectroscopy and x-ray diffraction, J. Appl. Phys. 109 (2011) 033502.

[44] H. Chen, Y.K. Li, C.S. Peng, H.F. Liu, Y.L. Liu, Q. Huang, et al., Crosshatching on a SiGe film grown on a Si(001) substrate studied by Raman mapping and atomic force microscopy, Phys. Rev. B 65 (2002) 233303.

[45] M.I. Alonso, K. Winer, Raman spectra of $c-Si1-xGex$ alloys, Phys. Rev. B 39 (1989) 10056.

[46] J.C. Tsang, P.M. Mooney, F. Dacol, J.O. Chu, Measurements of alloy composition and strain in thin $GexSi1-x$ layers, J. Appl. Phys. 75 (1994) 8098.

[47] J. Groenen, R. Carles, S. Christiansen, M. Albrecht, W. Dorsch, H.P. Strunk, et al., Phonons as probes in self-organized SiGe islands, Appl. Phys. Lett. 71 (1997) 3856.

[48] J. Li, A. Domenicucci, D. Chidambarrao, B. Greene, N. Rovdedo, J. Holt, et al., Stress and strain measurements in semiconductor device channel areas by convergent beam electron diffraction, Mater. Res. Soc. Symp. Proc. 913 (2006)Materials Research Society, 0913-D05-03.

[49] G.L. Wang, M. Moeen, A. Abedin, M. Kolahdouz, J. Luo, C.L. Qin, et al., Optimization of SiGe selective epitaxy for source/drain engineering in 22 nm node complementary metal—oxide semiconductor (CMOS), J. Appl. Phys. 114 (2013) 123511.

[50] J. Kouvetakis, J. Menendez, J. Tolle, Advanced Si-based semiconductors for energy and photonic applications, Diffusion and Defect Data Pt.B: Solid State Phenomena, 156—158 (2009) 77.

[51] J. Kouvetakis, J. Mathews, R. Roucka, A.V.G. Chizmeshya, J. Tolle, J. Menendez, Practical Materials Chemistry Approaches for Tuning Optical and Structural Properties of Group IV Semiconductors and Prototype Photonic Devices, IEEE Photonics J. 2 (2010) 924.

[52] O. Gurdal, P. Desjardins, J.R.A. Carlsson, N. Taylor, H.H. Radamson, J.-E. Sundgren, J.-E. Greene, Low-temperature growth and critical epitaxial thicknesses of fully strained metastable $Ge_{1-x}Sn_x$ ($x < = 0.26$) alloys on Ge(001)2×1, J. Appl. Phys. 83 (1998) 162.

[53] B. Vincent, F. Gencarelli, H. Bender, C. Merckling, B. Douhard, D.H. Petersen, O. Hensen, H.H. Henrichsen, J. Meersschaut, W. Vandervorst, M. Heyns, R. Loo, M. Caymax, Undoped and in-situ B doped GeSn epitaxial growth on Ge by atmospheric pressure-chemical vapor deposition, Appl. Phys. Lett. 99 (2011) 152103.

[54] H.H. Radamson, M. Noroozi, A. Jamshidi, M. Östling, Strain engineering in GeSnSi materials, ECS Trans. 50(9) (2013) 527.

[55] M.M. Naiini, H.H. Radamson, G.B. Malm, M. Östling, Embedded germanium PIN photodetectors in ALD slot waveguides, in: Ultimate Integration on Silicon (ULIS), 2014 15th International Conference Stockholm (2014) 45—48.

[56] M. Kolahdouz, P.T.Z. Adibi, A.A. Farniya, M. Shayestehaminzadeh, E. Trybom, L. Di Benedetto, et al., Selective growth of B- and C-doped SiGe layers in unprocessed and recessed Si openings for p-type metal—oxide—semiconductor field-effect transistors application, J. Electrochem. Soc. 157 (2010) H633.

[57] R. Loo, M. Caymax, Avoiding loading effects and facet growth key parameters for a successful implementation of selective epitaxial SiGe deposition for HBT-BiCMOS and high-mobility hetero-channel pMOS devices, Appl. Surf. Sci. 224 (2004) 24—30.

[58] S. Bodnar, E. de Berranger, P. Bouillon, M. Mouis, T. Skotnicki, J.L. Regolini, Selective Si and SiGe epitaxial heterostructures grown using an industrial low-pressure chemical vapor deposition module, J. Vac. Sci. Technol. B. Microelectron. Nanometer Struct. 15 (1997) 712.

[59] M. Kolahdouz, L. Maresca, R. Ghandi, A. Khatibi, H.H. Radamson, Kinetic model of SiGe selective epitaxial growth using RPCVD technique, J. Electrochem. Soc. 158 (2011) H457.

[60] G.L. Wang, A. Abedin, M. Moeen, M. Kolahdouz, J. Luo, T. Chen, H. H. Radamson, in press in Solid State Electronics (2014).

[61] M. Kolahdouz, L. Maresca, M. Ostling, D. Riley, R. Wise, H.H. Radamson, New method to calibrate the pattern dependency of selective epitaxy of SiGe layers, Solid State Electron. 53 (2009) 858.

[62] N. Tamura, Y. Shimamune, 45 nm CMOS technology with low temperature selective epitaxy of SiGe, Surf. Sci. 254 (2008) 6067.

[63] S. Mujumdar, K. Maitra, S. Datta, Layout-dependent strain optimization for p-channel trigate transistors, IEEE Trans. Electron Devices 59 (2012) 72—78.

[64] R. Chau, et al., Benchmarking nanotechnology for high-performance and low-power logic transistor applications, IEEE Trans. Nanotech. 4 (2005) 153.

[65] U.K. Mishra, P. Parikh, W. Yi-Feng, AlGaN/GaN HEMTs, an overview of device operation and applications, Proc. IEEE 90 (2002) 1022—1031.

[66] R.J. Trew, G.L. Bilbro, W. Kuang, Y. Liu, H. Yin, Microwave AlGaN/GaN HFETs, IEEE Microw. Mag. 6 (2005) 56–66.

[67] R. Chu, Z. Chen, P. Yi, S. Newman, S.P. Denbaars, U.K. Mishra, MOCVD-grown AlGaN buffer GaN HEMTs with V-gates for microwave power applications, IEEE Electron Device Lett. 30 (2009) 910–912.

[68] W. Saito, Y. Kakiuchi, T. Nitta, Y. Saito, T. Noda, H. Fujimoto, et al., Field-plate structure dependence of current collapse phenomena in high-voltage GaN-HEMTs, IEEE Trans. Microw. Theory Tech. 31 (2010) 659–661.

[69] K.S. Novoselov, A.K. Geim, S.V. Morozov, D. Jiang, Y. Zhang, S.V. Dubonos, et al., Electric field effect in atomically thin carbon films, Science 306 (2004) 666–669.

[70] A.K. Geim, K.S. Novoselov, The rise of graphene, Nat. Mater. 6 (2007) 183–191.

[71] M. Wilson, Electron band structure in graphene. Near K and K', the dispersion relation is relativistic with an effective mass being zer, Phys. Today (2006) 21–23.

[72] K.S. Novoselov, A.K. Geim, S.V. Morozov, D. Jiang, M.I. Katsnelson, I.V. Grigorieva, et al., Two-dimensional gas of massless Dirac fermions in graphene, Nature 438 (2005) 197–200.

[73] J.-H. Chen, C. Jang, S. Xiao, M. Ishigami, M.S. Fuhrer, Intrinsic and extrinsic performance limits of graphene devices on SiO2, Nat. Nanotechnol. 3 (2008) 206–209.

[74] C. Berger, Z. Song, X. Li, X. Wu, N. Brown, C. Naud, et al., Electronic confinement and coherence in patterned epitaxial grapheme, Science 312 (2006) 1191–1196.

[75] X. Du, I. Skachko, A. Barker, E.Y. Andrei, Approaching ballistic transport in suspended graphene, Nat. Nanotechnol. 3 (2008) 491–495.

[76] J. Moser, A. Barreiro, A. Bachtold, Current-induced cleaning of graphene, Appl. Phys. Lett. 91 (2007) 163513.

[77] Y. Xue, B. Wu, L. Jiang, Y. Guo, L. Huang, J. Chen, et al., Low temperature growth of highly nitrogen-doped single crystal graphene arrays by chemical vapor deposition, J. Am. Chem. Soc. 134 (2012) 11060–11063.

[78] K.K. Kim, A. Reina, Y. Shi, H. Park, L.-J. Li, Y.H. Lee, et al., Enhancing the conductivity of transparent graphene films via doping, Nanotechnology 21 (2010) 285205.

[79] P. Avouris, S. Member, M. Freitag, Graphene photonics, plasmonics, and optoelectronics, IEEE J. Sel. Top. Quantum Electron. 20 (2014) 6000112.

[80] P.K. Ang, A. Li, M. Jaiswal, Y. Wang, H.W. Hou, J.T.L. Thong, et al., Flow sensing of single cell by graphene transistor in a microfluidic channel, Nano Lett. 11 (2011) 5240–5246.

[81] G. Deligeorgis, M. Dragoman, D. Neculoiu, D. Dragoman, G. Konstantinidis, A. Cismaru, et al., Microwave switching of graphene field effect transistor at and far from the Dirac point, Appl. Phys. Lett. 96 (2010) 103105.

[82] Y.-M. Lin, C. Dimitrakopoulos, K.A. Jenkins, D.B. Farmer, H.-Y. Chiu, A. Grill, et al., 100-GHz transistors from wafer-scale epitaxial graphene, Science 327 (2010) 662.

[83] F. Schwierz, Graphene for electronic applications—transistors and more, in: 2010 IEEE Bipolar/BiCMOS Circuits Technol. Meet., 2010, pp. 173–179.

[84] F. Xia, D.B. Farmer, Y.-M. Lin, P. Avouris, Graphene field-effect transistors with high on/off current ratio and large transport band gap at room temperature, Nano Lett. 10 (2010) 715–718.

[85] E.V. Castro, K.S. Novoselov, S.V. Morozov, N.M.R. Peres, J.M.B. Lopes dos Santos, J. Nilsson, et al., Electronic properties of a biased graphene bilayer, J. Phys. Condens. Matter 22 (2010) 175503.

[86] K. Majumdar, K.V.R.M. Murali, N. Bhat, Y.-M. Lin, Intrinsic limits of subthreshold slope in biased bilayer graphene transistor, Appl. Phys. Lett. 96 (2010) 123504.

[87] C. Sealy, IBM demonstrates band gap and high on/off current in graphene FETs, Nano Today 5 (2010) 81.

[88] S. Nakaharai, T. Iijima, S. Ogawa, S.-L. Li, K. Tsukagoshi, S. Sato, et al., Current on—off operation of graphene transistor with dual gates and He ion irradiated channel, Phys. Status Solidi Vol. 10 (2013) 1608—1611.

[89] J. Hun Mun, B.J. Cho, Physical-gap-channel graphene field effect transistor with high on/ off current ratio for digital logic applications, Appl. Phys. Lett. 101 (2012) 143102.

[90] S. Vaziri, G. Lupina, C. Henkel, A.D. Smith, M. Ostling, J. Dabrowski, et al., A graphene-based hot electron transistor, Nano Lett. 13 (2013) 1435—1439.

[91] W. Li, S.-L. Li, K. Komatsu, A. Aparecido-Ferreira, Y.-F. Lin, Y. Xu, et al., Realization of graphene field-effect transistor with high-κ HCa2Nb3O10 nanoflake as top-gate dielectric, Appl. Phys. Lett. 103 (2013) 023113.

[92] J. Liu, Q. Qian, Y. Zou, G. Li, Y. Jin, K. Jiang, et al., Enhanced performance of graphene transistor with ion-gel top gate, Carbon 68 (2014) 480—486.

[93] R.R. Nair, P. Blake, A.N. Grigorenko, K.S. Novoselov, T.J. Booth, T. Stauber, et al., Fine structure constant defines visual transparency of graphene, Science 320 (2008) 1308.

[94] K.F. Mak, J. Shan, T.F. Heinz, Seeing many-body effects in single- and few-layer graphene: observation of two-dimensional saddle-point excitons, Phys. Rev. Lett. 106 (2011) 046401.

[95] L. Meng, Y. Wang, L. Zhang, S. Du, R. Wu, L. Li, et al., Buckled silicene formation on Ir (111), Nano Lett. 13 (2013) 685—690.

[96] A. Fleurence, R. Friedlein, T. Ozaki, H. Kawai, Y. Wang, Y. Yamada-Takamura, Experimental evidence for epitaxial silicene on diboride thin films, Phys. Rev. Lett. 108 (2012) 245501.

[97] D. Jose, A. Datta, Structures and Chemical Properties of Silicene: Unlike Graphene, Acc. Chem. Res. 47 (2014) 59.

[98] M. Ezawa, A topological insulator and helical zero mode in silicene under an inhomogeneous electric field, New J. Phys. 14 (2012) 033003.

[99] J. Gao, J. Zhao, Initial geometries, interaction mechanism and high stability of silicene on Ag(111) surface, Sci. Rep. 2 (2012) 861.

[100] L. Chen, C.-C. Liu, B. Feng, X. He, P. Cheng, Z. Ding, et al., Evidence for Dirac fermions in a honeycomb lattice based on silicon, Phys. Rev. Lett. 109 (2012) 056804.

[101] Z.-X. Guo, S. Furuya, J. Iwata, A. Oshiyama, Absence of Dirac electrons in silicene on Ag (111) surfaces, J. Phys. Soc. Jpn. 82 (2013) 063714.

[102] F. Bechstedt, L. Matthes, P. Gori, O. Pulci, Infrared absorbance of silicene and germanene, Appl. Phys. Lett. 100 (2012) 261906.

[103] M. Houssa, E. Scalise, K. Sankaran, G. Pourtois, V.V. Afanas'ev, A. Stesmans, Electronic properties of hydrogenated silicene and germanene, Appl. Phys. Lett. 98 (2011) 223107.

[104] S. Chang, L.T.L. Lee, T. Chen, (Chapter 17: Plasmon-Enhanced Excitonic Solar Cells) in: X. Wang, Z.M. Wang (Eds.), High-Efficiency Solar Cells: Physics, Materials, and Devices, Springer International Publishing, Switzerland, 2014, pp. 515—544.

[105] E. Simoen, G. Eneman, M. Bargallo Gonzalez, D. Kobayashi, A. Luque Rodriguez, J.-A. Jiménez Tejada, et al., High Doping Density/High Electric Field, Stress and Heterojunction Effects on the Characteristics of CMOS Compatible p-n Junctions, J. Electrochem. Soc. 158 (2011) R27—R36.

[106] E. Simoen, J. Mitard, G. Hellings, G. Eneman, B. DeJaeger, L. Witters, et al., Challenges and opportunities in advanced Ge pMOSFETs, Mater. Sci. Semiconductor Process. 15 (2012) 588−600.

[107] D.S. Yu, A. Chin, C.C. Liao, C.F. Lee, C.F. Cheng, M.F. Li, et al., Three-dimensional metal gate-high-κ-GOI CMOSFETs on 1-poly-6-metal 0.18-μm Si devices, IEEE Electron. Device Lett. 26 (2005) 118−120.

[108] A. Chin, W.B. Chen, B.S. Shie, K.C. Hsu, P.C. Chen, C.H. Cheng, et al., Metal-gate/high-κ CMOS scaling from Si to Ge at small EOT, Proc. 10th Int. Conf. Solid- State Integr. Circuit Technol. (ICSICT) (The IEEE, N. Y. (2010) 978−981.

[109] S. Dissanayake, K. Tomiyama, S. Sugahara, M. Takenaka, S. Takagi, High performance ultrathin (110)-oriented Ge-on-insulator p-channel metal-oxide-semiconductor field-effect transistors fabricated by Ge condensation technique, Appl. Phys. Express 3 (2010)041302/1−3.

[110] M. Kobayashi, J. Mitard, T. Irisawa, T.-Y. Hoffmann, M. Meuris, K. Saraswat, et al., On the High-field Transport and Uniaxial Stress Effect in Ge PFETs, IEEE Trans. Electron. Dev. 58 (2011) 384−391.

[111] Y. Shimura, T. Asano, O. Nakatsuka, S. Zaima, Crystallinity improvement of epitaxial Ge grown on a Ge(110) substrate by incorporation of Sn, Appl. Phys. Express 5 (2012)015501/ 1−3.

[112] S. Gupta, R. Chen, B. Magyari-Kope, H. Lin, B. Yang, A. Nainani, GeSn technology: extending the Ge electronics roadmap, IEDM Tech. Dig. (IEEE, N. Y.) (2011) 398−401.

[113] G. Han, S. Su, C. Zhan, Q. Zhou, Y. Yang, L. Wang, et al., High-mobility germanium-tin (GeSn) P-channel MOSFETs featuring metallic source/drain and sub-370C process modules, IEDM Tech. Dig. (IEEE, N. Y.) (2011) 402−404.

[114] M.L. Lee, E.A. Fitzgerald, M.T. Bulsara, M.T. Currie, A. Lochtefeld, Strained Si, SiGe, and Ge channels for high-mobility metal-oxide semiconductor field-effect transistors, J. Appl. Phys. 97 (2005) 011101.

[115] A. Delabie, F. Bellenger, M. Houssa, T. Conard, S. Van Elshocht, M. Caymax, et al., Effective electrical passivation of Ge (100) for high-k gate dielectric layers using germanium oxide, Appl. Phys. Lett. 91 (2007) 141.

[116] A. Toriumi, T. Tabata, C.-H. Lee, T. Nishimura, K. Kita, K. Nagashio, Opportunities and challenges for Ge CMOS − Control of interfacing field on Ge is a key, Microelectronic Eng. 86 (2009) 1571−1576.

[117] A. Toriumi, C.H. Lee, S.K. Wang, T. Tabata, M. Yoshida, D.D. Zhao, et al., Material potential and scalability challenges of germanium CMOS, Int. Electron Devices Meeting - IEDM (2011) 28.4.1−28.4.4.

[118] M. Kobayashi, G. Thareja, M. Ishibashi, Y. Sun, P. Griffin, J. McVittie, Radical oxidation of germanium for interface gate dielectric GeO2 formation in metal-insulator-semiconductor gate stack, J. Appl. Phys. 106 (2009) 104117.

[119] D. Kuzum, J.-H. Park, T. Krishnamohan, H. S-. P. Wong, K.C. Saraswat, The effect of donor/acceptor nature of interface traps on Ge MOSFET characteristics, IEEE Trans. Electron. Devices 58 (2011) 1015−1022.

[120] M. Kobayashi, G. Thareja, M. Ishibashi, Y. Sun, P. Griffin, J. McVittie, Radical oxidation of germanium for interface gate dielectric GeO2 formation in metal-insulator-semiconductor gate stack, J. Appl. Phys. 106 (2009) 104117.

[121] D.H. Lee, H. Imajo, T. Kanashima, M. Okuyama, Improvement in the property of field effect transistor having the HfO2/Ge structure fabricated by photoassisted metal organic chemical vapor deposition with fluorine treatment, Jpn. J. Appl. Phys. 50 (2011) 04DA11.

[122] S. Sioncke, H.C. Lin, G. Brammertz, A. Delabie, T. Connard, A. Franquet, et al., S-passivation of the Ge gate stack: tuning the gate stack properties by changing the atomic layer deposition oxidant precursor, J. Electrochem. Soc. 158 (2011) H687−H692.

[123] A. Dimoulas, Y. Panaiotatos, A. Sotiropoulos, P. Tsipas, D.P. Brunco, G. Nicholas, Germanium FETs and capacitors with rare earth CeO_2/HfO_2 gates, Solid-State Electron. 51 (2007) 1508−1514.

[124] C. Henkel, S. Abermann, O. Bethge, G. Pozzovivo, P. Klang, M. Reiche, et al., Ge p-MOSFETs With Scaled ALD La_2O_3/ZrO_2 Gate Dielectrics, IEEE Trans. Electron. Devices 57 (2010) 3295−3302.

[125] C. Andersson, M.J. Suess, D.J. Webb, C. Marchiori, M. Sousa, D. Caimi, et al., Mobility and D_{it} distributions for p-channel MOSFETs with $HfO_2/LaGeOx$ passivating layers on germanium, J. Appl. Phys. 110 (2011)114503/1−7.

[126] T. Krishnamohan, D. Kim, S. Raghunathan, K. Saraswat, Double-Gate strained Ge heterostructure tunnelingFET (TFET) with record high drive currents $\ll 60$mV/dec subthreshold slope mV/dec subthreshold slope, electron devices meeting, IEDM, 2008, IEEE international electron devices meeting, IEDM (2008) 1.

Basics of Integrated Photonics

GENERAL

Integrated photonics devices, and indeed most photonics systems, include sources (lasers, LEDs), light detectors, and an optical waveguide (or possibly free space) based "fabric" or network in between to transport light in some shape. In the waveguided version, this fabric can include optical modulators, changing amplitude, phase, and/or polarization of the light,

Monolithic Nanoscale Photonics—Electronics Integration in Silicon and Other Group IV Elements.
DOI: http://dx.doi.org/10.1016/B978-0-12-419975-0.00002-7
© 2015 Elsevier Ltd. All rights reserved.

as well as switches, to redirect light, optical amplifiers, and wavelength selective structures for filtering, wavelength multiplexing and demultiplexing, and other operations involving wavelengths or light frequency. Integrated photonics has developed at a considerably slower pace than integrated electronics, as a matter of fact, it was a subject of a joke stating that "integrated photonics is the technology of the future and will remain the technology of the future." However, this state of affairs has altogether been changed by progress in material technology in III−V compounds (GaAs, InP systems, etc), ferroelectrics (LiNbO$_3$), silicon, polymers, and metal optics.

The basic structure of an integrated photonics circuit is the optical waveguide (Figure 2.1). In most photonics integration applications and in all such applications where highest performance is sought, these waveguides are **single mode**, by which is meant that only one spatial mode can propagate, and other modes are evanescent or cut off [1]. However, the waveguides are not strictly single mode and normally support two orthogonal polarizations, a fact that has caused a number of problems in the past and present. The reason is that the standard single mode fiber does not preserve light polarization, and hence photonic elements in a fabric of such fibers need to work independently of the state of polarization of the input light. Such polarization independence normally means that compromises in the device performance have to be made.

A so-called channel waveguide confines light in two dimensions (in the so-called core) while it propagates in the third dimension. In the more exotic so-called plasmonic waveguide, light can be guided along a

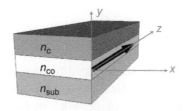

Figure 2.1 Schematic diagram of a planar dielectric waveguide in Cartesian coordinate system. n$_c$, n$_{co}$, and n$_{sub}$ stand for refractive indices of the cover layer, guided wave layer (core), and substrate, respectively. Light is confined in the y direction and generally most of the optical field resides in the core. Light propagation is in the z direction.

single plasmonic, usually metal-dielectric interface, as will be briefly discussed below. The confinement of light in the two dimensions orthogonal to the direction of light propagation is accomplished by *total internal reflection* [1], just like in an optical fiber, by having a core with higher refractive index than the surrounding. This surrounding is called cladding in a fiber and is in general partly the substrate in a PIC.

Figure 2.2 shows some basic structures of integrated photonic waveguides, with a central core of higher refractive index than the surrounding medium, cladding, or substrate. The optical field is also shown in Figure 2.3.

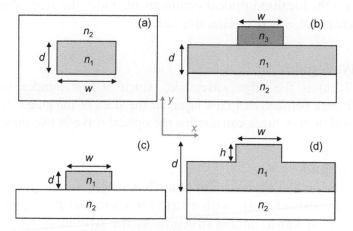

Figure 2.2 Nonplanar dielectric waveguide types: (a) buried channel, (b) strip-loaded, (c) ridge, and (d) rib.

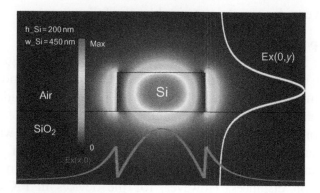

Figure 2.3 Electric field distribution of Transverse Electric (TE) mode in a silicon channel dielectric waveguide, the Ex(0, y) and Ex(x, 0) curves express the amplitude distribution in x- and y-axis directions, respectively; the substrate material is SiO₂, and the cover is air. The guided layer is made of silicon material with the geometry parameters are height = 200 nm and width = 450 nm.

Buried Channel Waveguide

Figure 2.2(a) shows a buried channel waveguide, and it consists of a high-index waveguiding core buried in a low-index cladding. The optical wave can be confined in two dimensions due to differences of refractive index between the core and the cladding.

Strip-Loaded Waveguide

Figure 2.2(b) is the geometry of a strip-loaded waveguide, which is composed of three dielectric layers: a substrate, a planar layer, and then a ridge. The planar waveguide (without the strip) already provides optical confinement in the vertical direction (y-axis), and the additional strip can offer localized optical confinement under the strip, due to the local increase of effective refractive index.

Ridge Waveguide

Figure 2.2(c) is the ridge waveguide, which is a step-index structure. The difference between dielectric layers at the sides of the guide, as well as the top and bottom faces, can confine the optical wave in two dimensions.

Rib Waveguide

Figure 2.2(d) is the cross-section of a rib waveguide. The guiding layer basically consists of a slab with a strip (or several strips) superimposed onto it, which has a similar structure as the strip-loaded waveguide, and the strip is part of the waveguiding core.

The waveguides are characterized by the following:

- Optical power loss, usually in dB/cm.
- Effective index, usually denoted by N or N_{eff}, which is equal to β/k_0, where β is the real part of the propagation constant and k_0 is the wave number in vacuum. The effective index is, for guided waves, larger than cladding index but smaller than core index.
- Dispersion, i.e. the variation of the effective index with wavelength. This determines limitations in the propagation length of very short pulses but is normally not so important in PICs due to the small propagation distances. However, for the devices in PICs, such as filters, the so-called group delay dispersion, i.e. the derivative of the group delay with respect to angular frequency can be significant and important.
- Geometrical waveguide and optical field cross-sectional area.

- The useful wavelength range for light transportation. These are characterized by several "bands" between 1260 and 1675 nm for ICT applications.

Waveguide parameters and propagation characteristics for waveguides fabricated with different material compositions are presented in Table 2.1. SOI is silicon on insulator, usually quartz (SiO_2). Small waveguide bending radii are desirable for dense integration.

The waveguides connect different functional elements—lasers, modulators, switches, optical amplifiers, wavelength selective devices, detectors, etc.—and are generally also used to create these device structures, as will be described in the following section. Figure 2.4

Table 2.1 Waveguide Parameters for Different Materials						
Column	1	2	3	4	5	6
Characteristics	SiO_2 Low Δ	SiO_2 Medium Δ	SiO_2 High Δ	$SiON_x$	III/V	SOI
Index difference Δ (%) $\Delta = \dfrac{n_{core} - n_{clad}}{n_{core}}$	0.3	0.45	0.75	3.3	7.0 (46)	41 (46)
Core size (μm)	8 × 8	7 × 7	6 × 6	3 × 2	2.5 × 0.5 (0.2 × 0.5)	0.2 × 0.5 0.3 × 0.3
Loss (dB/cm)	<0.01	0.02	0.04	0.1	2.5–3.5	1.8–2.0
Coupling loss (dB/point)	<0.1	0.1	0.4	3.7 (2)	5	6.8 (0.8)
Waveguide bending radius (mm)	25	15	5	0.8	0.25 (0.005)	0.002–0.005

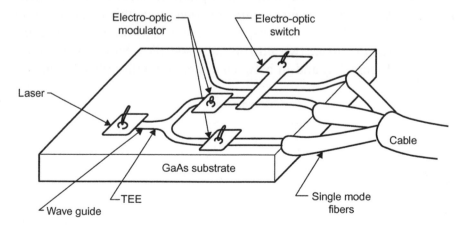

Figure 2.4 Artists sketch of a monolithic integrated photonics circuit, encompassing single mode waveguides on a planar surface. The waveguides connect the laser source to modulators (see below) and a switch which can direct light to either of two output ports. All three outputs are connected to a single mode fiber [2].

shows one of the first publications introducing the concept of integrated photonics.

In general, it is desired to have a waveguide and optical field with small cross-sectional area since this corresponds to small area on the devices. This is of importance for integration as well as for low power consumption, an issue we will return to later. Here silicon plays an important role since it has the largest practically useable refractive index, around 3.5 at 1500 nm wavelength. This allows much more compact device structures than what was previously allowed. On the other hand, still smaller structures are desired, an issue treated in the section on plasmonic/silicon devices.

BASICS OF LASERS, MODULATORS, DETECTORS, AND WAVELENGTH SELECTIVE DEVICES

Lasers

There is an abundance of literature on lasers in different materials. As of today the prevalent laser types are III−V semiconductor lasers [3], fiber lasers [4], and different solid state lasers. Here we only describe the very basic characteristics of a laser; the heterostructure semiconductor laser will be described in Chapter 3 in conjunction with treating the research on group IV−based lasers. The semiconductor heterostructure was awarded the Nobel Prize in 2000 (Kroemer and Alferov).

Laser stands for *light amplification by stimulated emission of radiation*; however, it should be more appropriate to label it "loser": *light oscillation by stimulated emission of radiation*, since this is the accurate description. For obvious reasons, this did not happen. The laser comprises a resonator (e.g., two mirrors in a so-called Fabry−Perot (FP) resonator, Figure 2.5) and in between a medium with optical gain, i.e.,

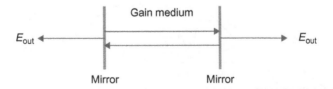

*Figure 2.5 Principle of operation of a laser in the Fabry−Perot cavity configuration: Spontaneous of the gain medium is partially reflected by the mirrors such that light is trapped between the mirrors. Light loss through the mirrors is compensated by the gain of the gain medium, such that unity amplification is resulted after one full roundtrip in the cavity; this is the lasing condition eq. (FP). The output power is proportional to the square of the output light electric field E*out*.*

the capability to amplify light. A simple way to describe the gain medium is to model it as a two-level system (atom, ion, semiconductor) with the electron in the ground or in the excited state; see [5] for further details. In the ground state, the system will absorb an incident photon, being excited to the higher energy state. For the electron in the excited state, two things can happen: Either, in the absence of any energy light field in the resonator, the system can decay (carrier recombination) to the lower state by the so-called spontaneous emission (coupling to all possible photon states being "populated" by the so-called zero field fluctuations [6] surrounding the system) or by the so-called stimulated emission, where an incident photon gives rise to two outgoing photons (in the same photonic mode [1], giving rise to amplification). In order to provide amplification, the majority of the two-level systems have to be in the excited state, what is called *inversion*, and are brought there by pumping, either with light or, in semiconductor lasers, by electric current injection, a more practical method.

It should be noted that in order to accomplish and maintain inversion, a system with three or more levels is required, where an uppermost state is pumped and rapidly generally nonradiatively relaxes to a lower state, from which the transition to the ground state brings about the emission of the desired photon. This relaxation from the uppermost state has to be much faster than the subsequent relaxation with photon emission.

When the material gain in the cavity is increased such that it (nearly) equals the losses (by light scattering and absorption in the resonator waveguide and light power losses through the mirrors), the gain cannot be increased any more, and the gain is what is called "clamped" to a threshold value, otherwise an unphysical power runaway due to infinite amplification would occur. Above threshold, all increases in the pump power will (ideally), by radiative recombination, be turned into increases in light power, since further increase in inversion is not possible, as noted. However, the gain will always (except in some exotic laser structures) be somewhat smaller than the losses, the balance being the spontaneous emission: An excited two-level system can, as noted, in addition to decay by stimulated emission by an incident photon in the lasing mode in question decay by spontaneous emission i.e., coupling of the two-level system to all available photonics modes as noted above (i.e., not only the lasing mode), in contrast to the stimulated emission which is coupled to the lasing mode. The spontaneous emission rate becomes proportionally smaller with increasing power in the lasing mode.

A simple equation for the threshold condition of a Fabry—Perot cavity type laser (Figure 2.5) is

$$e^{(g-\alpha)L} R = 1 \tag{2.1}$$

where g is the net spatial gain coefficient (in cm^{-1}) of the laser material, such as GaAs with population inversion. α is analogously the spatial loss coefficient for losses unrelated to the two-level system, such as scattering in the cavity, L is the cavity length and R the mirror reflectance.

Key performance parameters for an electrically pumped laser are:

- Threshold current and current density: A typical value of threshold current is 10 mA (distributed feedback laser), while threshold current density is around 2 kA/cm^2.
- Slope efficiency: Optical output power (in mW) per input current (mA) above threshold
- Wall plug efficiency: Power output normalized to electrical power input
- Spectral linewidth
- Output light field noise, above quantum (Poisson) noise (for a discussion of noise in lasers refer to [7])
- Temperature properties (uncooled lasers are very desirable in many applications)
- Physical size

BASICS OF PHOTONIC DETECTORS

Photonic semiconductor detectors or optical sensors are devices which provide an electrical response when an electromagnetic radiation with energy $\hbar w$ is illuminated on the active region of detector and is absorbed if $\hbar w > E_g$, where E_g is the bandgap of the semiconductor material. No absorption occurs for $\hbar w < E_g$ and the light passes through the material.

The light absorption may generate electron—hole pairs. The structure of the detector is a p—n junction, p—i—n diode or a Schottky barrier diode (metal/semiconductor/metal or MSM). The diode is reverse biased and the electrons and holes are separated in the depletion region.

These structures which are sensitive to light with a certain wavelength can be considered as solar cells generating electricity or as photodetectors responding to the incident radiation.

Operating modes:

No light absorption $\hbar\omega < E_g$
Light absorption $\hbar\omega > E_g$ or $hc/\lambda > E_g$ (Figure 2.6)

The condition for light absorption is $hc/\lambda > E_g$ which leads to the definition of cutoff wavelength, λ_c:

$$\lambda_c = \frac{hc}{E_g} = \frac{1240}{E_g \ (\text{eV})} \ \text{nm} \tag{2.2}$$

The absorption coefficient and the cutoff frequency of different materials are the main operating factors of a detector. The absorption coefficient is a value which shows how far light with a certain wavelength can penetrate into a material until it is absorbed. If the absorption coefficient is low, then the light is sparsely absorbed, and the material

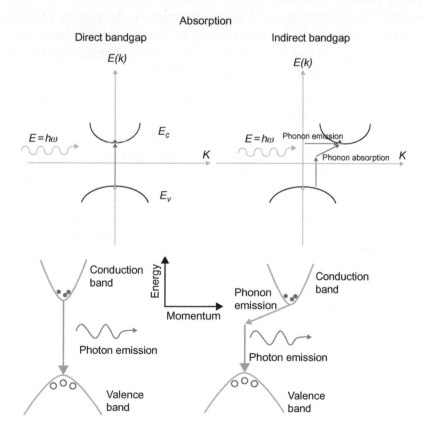

Figure 2.6 Light absorption of direct and indirect bandgap semiconductor.

will act as transparent to that wavelength. The absorption coefficient for direct bandgap semiconductors has a sharper edge compared to indirect semiconductors which contain phonon absorption or emission. The absorption coefficient for different semiconductors in range of 500–2000 nm is shown in Figure 2.7.

DETECTOR CHARACTERISTICS

A PIN detector is a diode which functions under a high-level injection. In this condition, a large flow of charge carriers comes to the intrinsic part from both the "p" and "n" regions. When the injected holes and electrons reach to an equilibrium point, the diode will begin to conduct current. In forward bias, the injected carrier concentration is usually some orders of magnitude greater than the intrinsic level. As a result of this high-level injection and the depletion region, an electric field is generated which extends over the whole length of the region. This electric field contributes to faster transport of charge carriers from p to n region of diode yielding quicker operation of the diode and making the device excellent for high-frequency application.

MSM: The detector operates when an electric voltage is applied to the metal electrodes. When light is illuminated to the area between the electrodes, it produces electrons and holes which are conducted by the established electric field and therefore can generate a photocurrent. MSM detectors can operate faster than photodiodes and their detection

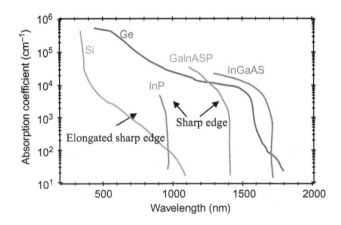

Figure 2.7 The absorption coefficient for several semiconductor materials [8].

bandwidths may extend hundreds of GHz which makes these devices a candidate for future optical fiber communications [9] (Figure 2.8).

The maximum current from a p−i−n detector with the intrinsic thickness of t_{in} where the light with the incident radiation intensity (in W/m^2) of P_1 is illuminated to the active area (with an area of A) is calculated from [10]:

$$I_{t_{in}} = qA \int_0^L G_{t_{in}} \, dx \tag{2.3}$$

where $G_{t_{in}}$ is the electron−hole pair generation rate which is defined as

$$G_{t_{in}} = \frac{P_1}{\hbar\omega} \alpha \exp(-\alpha x) \tag{2.4}$$

Then,

$$I_{t_{in}} = \frac{qAP_1(1 - \exp(\alpha L))}{\hbar\omega} \tag{2.5}$$

The maximum collection efficiency of detector when the light reflection from the detector surface is taken into account is:

$$Q_c = (1 - R)(1 - \exp(\alpha L)) \tag{2.6}$$

where R is the reflection coefficient for incident beam (Figure 2.9).

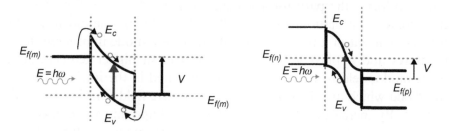

Figure 2.8 Illustration of how PIN and MSM operate when light with energy of $\hbar\omega$ may excite an electron into the conduction band.

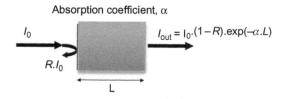

Figure 2.9 Illustration of light illumination, absorption, and transition through detector layer.

RESPONSIVITY

This is a design parameter which measures the electrical **output** per optical input (or input−output **gain**) for a detector. A relationship between the quantum efficiency, η, and responsivity, R_s, is given by

$$R_s = \frac{q\eta}{h\nu} = \frac{\eta\lambda\,(\text{nm})}{1240} \qquad (2.7)$$

R_s has typically a linear relation with wavelength but real photodetectors exhibit a deviation from the ideal behavior due to photogenerated carrier trapping [10].

DARK CURRENT

An ideal photodetector should respond strongly in the presence of radiation and give no electric output current in the absence of radiation (i.e., in darkness). However, a dark current is always present for practical photodetectors. It appears as a noise for their performance and a low dark current is thus a figure of merit for photodetectors. The main reason for the dark current is usually crystal defects in the detector material and imperfection in contacts or other active regions. As an example, Ge p−i−n diodes which are vastly used for infrared detection, it is shown that defect-induced traps in the forbidden energy bandgap forms the path for a leakage current which can flow even in the absence of radiation. Figure 2.10(a) shows the measurement results for dark current of a reversely biased Ge p−i−n diode in different temperatures and applied voltages to the diode. In higher temperatures, more random electron−hole pairs are generated in the intrinsic part of the diode which are then drifted by the electric field forming the dark current. By increasing the reverse applied voltage and the consequent band-bending (Figure 2.10(b)), the chance of tunneling through the bandgap traps increases, hence the dark current. The band-bending effect also puts a limit on the intrinsic region downscaling as the thinner the intrinsic region gets, the dark current also increases as a result of easier carrier tunneling [11].

NOISE CHARACTERISTICS OF PHOTODETECTORS

The lower limit of light detection is determined by the noise level of the detector [7,10,12].

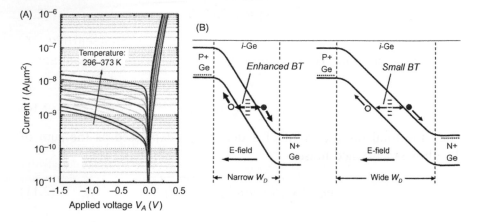

Figure 2.10 (a) Dark current measured in different temperatures for various applied reverse voltages to a Ge p–i–n diode and (b) band-bending in high applied voltages for two diodes with different intrinsic region widths [11].

The signal-to-noise ratio is finger a figure of merit which has to be high for a detector. The ratio is affected by the parasitic resistances, defect density in the detector structure and detector design.

When the applied voltage to a photodetector is in range of few mV, the dark current has a linear relationship with voltage. Shunt resistance (R_{sh}) is defined as the slope of the dark current versus applied voltage to the detector. R_{sh} is the cause of Johnson noise, n_J (or thermal noise). This type of noise relates to the random thermal fluctuations of the carriers and it is inversely proportional to the shunt resistance. Hence, in order to reduce the noise, a large resistance value is required and intended for a measurement setup [7]. The generated noise is obtained from

$$n_J = \sqrt{\frac{4kTB}{R_{sh}}} \tag{2.8}$$

where B is noise bandwidth, T stands for absolute temperature of diode, and k is Boltzmann's constant.

When the PIN detector is in reverse bias, the dark current, I_D, appears and the shot noise is created. Shot noise is associated to the randomness of the incident photon stream and is given by

$$n_S(I_D) = \sqrt{2qI_D B} \qquad (2.9)$$

When light is illuminated on the detector, the photocurrent, I_L will possess another type of shot noise which can be dominant if $I_L \gg I_D$, then

$$n_S(I_L) = \sqrt{2qI_L B} \qquad (2.10)$$

Then the total noise current of a detector is expressed as the sum of the above-mentioned noises:

$$n_{Sn} = \sqrt{n_J^2 + n_S(I_D)^2 + n_S(I_L)^2} \qquad (2.11)$$

MODULATORS: PRINCIPLES AND MECHANISMS OF OPTICAL MODULATION

Optical or photonic modulators are used in many applications, a prevalent one being in ICT systems, Figure 2.11. Amplitude (power), phase, and polarization modulation can be implemented, thus modulators can be seen as basic light field transformation devices. An input field

$$\hat{E} = \hat{x} A_x \sin(\omega t) \qquad (2.12)$$

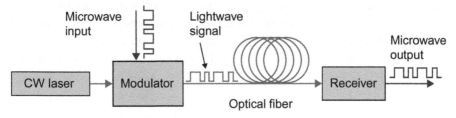

Important parameters:

Voltage swing V_{PP}
Modulation speed f_{3dB}
Extinction ratio
Insertion loss
Chirp parameter α_{mod}
Saturation power P_S

Figure 2.11 Representative use of an optical modulator in a fiber-optic commutation system. The continuous wave laser is coupled to the modulator waveguide(s) and the light is modulated as described above by the microwave input signal (intensity modulation in the example) and transmitted over the fiber to a receiver. It is also possible to modulate the current to the laser directly, in which case the modulator is superfluous. The use of an external modulator is in many cases required to generate adequate signal quality, e.g., for long distance transmission or closely spaced wavelength channels.

polarized in the x-direction with amplitude A_x and angular frequency ω can be transformed into

$$\hat{E} = \hat{x}A_x(t)\sin(\omega(t)t\varphi_1(t)) + \hat{y}A_y(t)\sin(\omega(t)t + \varphi_2(t)) \qquad (2.13)$$

where amplitude, frequency, phase, and polarization modulation are present. Depending on the physical mechanism used, these different quantities are more or less related to one another; obviously frequency and phase are related via a time derivative. This transformation operation in essence rests on, for example, using an electronic signal to controllably change the complex refractive index $\tilde{n} = n' + in''$ of the modulator waveguide structure, where $\tilde{n} = \sqrt{\tilde{\varepsilon}\tilde{\mu}}$, and $\tilde{\varepsilon}$ is the complex dielectric constant. The real and imaginary parts of the complex dielectric constant are connected by the Kramers–Krönig relations [12]. The magnetic permeability μ is taken as 1 for common cases. The change of the complex refractive index can be achieved by a wide variety of materials and employing a plethora of physical effects. Variation of the imaginary part of the refractive index gives intensity modulation: absorption or amplification (often accompanied by an undesired phase modulation), whereas modulation of the real part of the refractive index (phase modulation) is the most versatile and can in principle be employed to modulate all the parameters above. Several physical effects can be employed here, of which one of the most important ones is the Pockels' or electro-optic effect [12]. The refractive index change Δn using the Pockels' effect is customarily written as

$$\Delta n = -(1/2)n^3 rE \qquad (2.14)$$

where n is the refractive index of the material employed, such as $LiNbO_3$ or an electro-optic polymer, r is the appropriate element of the electro-optic tensor [12], usually r_{33} is quoted, which is around 30 pm/V in $LiNbO_3$. E is the applied electric RF field. The induced index changes are in the order of 10^{-4} to $>10^{-2}$ depending on electric RF field strengths and materials.

As an example, by changing the phase in interferometer-type modulators, such as the important Mach–Zehnder modulators (Figure 2.12), one can modulate amplitude or intensity. By changing the imaginary part of the refractive index, absorption changes are resulted, due to, for example, the Franz–Keldysh or the quantum confined stark effects [13].

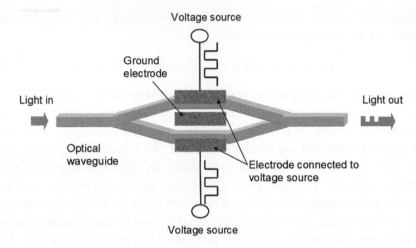

Figure 2.12 Schematic of a Mach–Zehnder modulator: Light is at the left port coupled into a single mode waveguide and split equally into two branches. In case the arms are of equal lengths and with no voltages applied over either arm, the light in the two waveguides will hit the right merging point in phase and be coupled to the output waveguide with no power loss. In case the combined or single action of the modulating signals effect a π phase shift between the signals when they arrive at the merging point, they will be out of phase, interfere destructively, and radiate out of the output waveguide, such that there is little power output in the output waveguide. The out-of-phase signals effectively constitute the next higher order (odd) mode, and since all waveguides are single mode, they do not "fit" into the output waveguide and are radiated away.

With reference to polarization, it should be mentioned that basically all modulators will only operate efficiently for a single polarization. In, for example, LiNbO₃ modulators, based on employing the electro-optic effect, the polarization dependence is due to the anisotropy of the material. In semiconductor modulators, the matrix elements of the involved transitions are different for different polarizations. For modulators, the restriction to single polarization operation is not overly severe (in contrast to switches), since in most cases, the modulator is close to or integrated with a light source. Polarization-independent modulators will require special and challenging designs which will in general compromise performance.

Key performance parameters for modulators are:

- Insertion loss and extinction ratio: Should generally be lower than a few dB and larger than 10 dB respectively.
- Drive power (electrical drive power).
- Chirping, i.e., wavelength changes with changing power output.
- Bandwidth (phase and amplitude): Largest reported values are in excess of 100 GHz.
- Return loss (electrical and optical).

- Wavelength range: It is an advantage if the modulator can operate over a reasonably large wavelength range with no other modification than the drive voltage. The latter is true for $LiNbO_3$ modulators, but not for, for example, electroabsorption modulators.
- Physical size: Many applications would favor sizes smaller than a few hundred microns.
- Integratability with light sources.

PHOTONICS SWITCHES: SPATIAL ROUTING OF HIGH-SPEED DATA STREAMS

Switches

Figure 2.11 shows a basic optical transmission system. However, in today's complex photonics systems, there are multiple transmitters and receivers, interconnected in what is called a network fabric. This fabric requires more functionality than only transmitting and receiving, and important functions are electronically controllable spatial routing of information as well as operating on a multitude of wavelength channels. Here we are, in addition to simple filtering, talking about adding and dropping wavelengths, multiplexing, or demultiplexing of wavelength bands, wavelength conversion, etc. Some of the functions and devices for operation of wavelengths are treated in the next section. This section, however, deals with one of the key devices, i.e., the optical space switch, in its simplest configuration depicted in Figure 2.13.

This 2×2 switch in Figure 2.13 is made up of a directional coupler [1], with two inputs (lower left, only one fiber coupled to the chip) and two outputs. Without any applied voltage to the electrodes, the coupler can be configured such that input light can couple over totally to the adjacent waveguide via evanescent field coupling, effecting the so-called cross state, as shown in the figure. Input light in the other waveguide will likewise couple over to the opposite waveguide, and the device can consequently handle two independent signals. Applying a voltage over the structure will generate an electric field via, for example, the Pockels effect, which will change the refractive index of, for example, one of the waveguides. This will cause the waveguides to have different propagation constants (being mismatched as the term is) and that can be arranged to inhibit coupling, such that light stays in the input arm (*bar state*). The physics here is analogous to trying to couple two oscillators, where the cross state corresponds to the periodic coupling of energy between

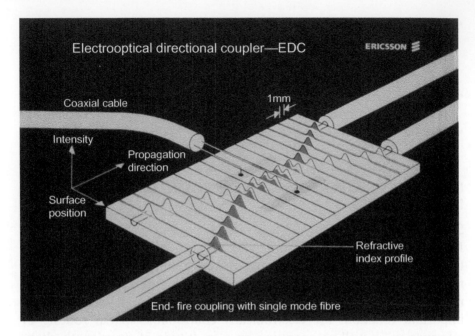

Figure 2.13 Schematic of an optical space switch, implemented as a 2 × 2 s directional coupler with two inputs (lower left), one of which is coupled to a fiber and two outputs. Without any applied voltage to the electrodes, the coupler can be configured to couple input light to the adjacent waveguide, implementing the so-called cross state, as shown in the figure. Applying a voltage over the structure will generate an electric field via, for example, the Pockels effect, which will change the refractive index of one of the waveguides causing the waveguides to have different propagation constants which will inhibit coupling, such that light stays in the input arm (bar state). Note: The electrodes should actually be arranged above the waveguides, delineated by the waveguide refractive index profiles.

frequency matched oscillators, and the bar state is achieved by sufficient frequency mismatch of the oscillators, inhibiting coupling such that energy stays in the oscillator which is excited.

Key performance parameters for space switches are:

- Insertion loss (as low as possible).
- Crosstalk (i.e., power remaining in the undesired output).
- Optical bandwidth: How wide a wavelengths range can the device handle: Very wide bandwidth switches can switch ingress information regardless of signal speed, modulation format, digital or analog formats, and are thus called *transparent*, a very important feature for optical networks.
- Switching speed, i.e., speed of rearrangement from, for example, cross to bar state, where the requirements very much depend on the application.

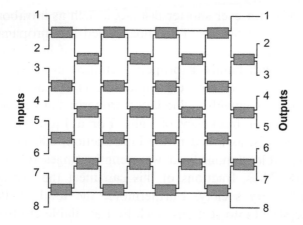

Figure 2.14 A rearrangeably nonblocking 8 × 8 switch, made up of 2 × 2 switches. Rearrangeably nonblocking implies that a free output port can only be reached with certainty if existing confections are rearranged, which is a disadvantage in some applications.

- Physical size.
- Polarization sensitivity, i.e., the ability to handle both polarizations.

An important feature of space switches is the possibility to concatenate them into switch arrays for deployment in photonics networks. Such space switched networks had already been investigated in detail during the last century for electronic switching systems in the emerging telephone systems. A comprehensive treatment of optical space switch arrays can be found in Ref. [14]. Figure 2.14 shows an example of an 8 × 8 switch array made up of 2 × 2 switches. An important property of switch array is their blocking properties: An array where a free output port can be reached without rearranging existing connections through the array is called *strictly nonblocking*, otherwise we have a rearrangably nonblocking switch array. The switch of Figure 2.14 is only rearrangeably nonblocking.

Switch arrays used as transparent fabrics are sometimes called cross connects or circuit switches.

Devices for Wavelength Division Multiplexed Systems

The great and ubiquitous breakthrough and deployment of the guided wave optical media for transmission over extreme distances and lately the consideration for ever shorter distances, such as intraboard, can be attributed to several factors: The unparalleled low propagation losses in the optical waveguide, virtually independent of light wavelength in relevant wavelength bands, a property retained even when shrinking waveguide cross-section to order of microns, as well as the giant available information bandwidth. The latter property can be understood in terms of the carrier frequency of light: For a 1 μm wavelength, the carrier frequency is, as noted in the introduction, 300 THz. The commercially available standardized wavelength ranges (O to U bands) cover 60 THz. Thus, fractions of this can meet the demands of our information hungry society. Furthermore, the total available bandwidth can be sliced into sections, such that available electronics equipment can cope with it, for example, using slices of several 100 GBaud. This slicing into frequency sections for various applications needs the corresponding wavelength selective or routing devices, i.e., devices for wavelength filtering and wavelength (de)multiplexing, key elements in photonics systems, usually named wavelength division multiplexed (WDM) devices.

Key performance parameters for WDM devices are:

- Spectral bandwidth and passband shape (see Figure 2.15). The latter is important for concatenation of devices.
- Channel density or spacing: Coarse WDM, typically 20 nm spacing, dense WDM, typically 0.8 nm spacing.
- Crosstalk, i.e., the influence of adjacent channels, measured in dB.
- Insertion loss in dB.
- Size.
- Thermal properties, very important for, for example, channels packed in the standard 100 GHz channel separation grid.
- Number of channels supported.
- Tunability, i.e., the capability to change the passband center frequency, usually by changing the refractive index of some part of the device, for example, by employing the Pockels effect.
- Polarization sensitivity, i.e., the ability to handle both polarizations.

A common standard for the separation of optical channels is 100 GHz, corresponding to 0.3 nm at a wavelength of 1000 or 0.8 nm at a standard telecom wavelength of 1550 nm.

Most or all of these WDM devices are based on optical interference (see the next section). We realize immediately that we in most cases have to deal with the issue of polarization sensitivity, since interference occurs only when polarizations are the same.

Devices Based on Spectrally Dependent Interference Effects

- Periodic refractive index changes in fibers, waveguides, etc. These are the well-known Bragg filters.
- Mach–Zehnder structures: See above where by having different "arm" lengths, a filter with periodic sinusoidal response is resulted but for filter concatenation it is not very optimal.
- Arrayed waveguide gratings [15].
- Ring resonators.
- Interference filters: Waveguided version of the same type of filters which serve as Autoregressive filters (AR) on camera lenses.
- Couplers (see previous section) (Figure 2.15).

It is also possible to combine space switches and wavelength selective functions into fabrics called WXC: wavelength cross connects (see Figure 2.16).

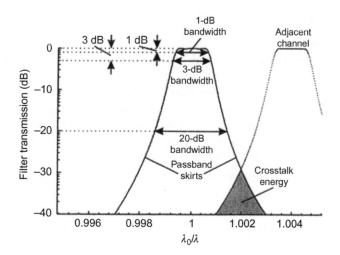

Figure 2.15 Some key properties of photonic filters: bandwidth and cross talk. The latter should be as low as possible, say −20 dB. The former can be engineered to desired values. However, equally important is the shape of the passband: a flat so-called top hat feature is highly desirable for concatenation purposes. For example, a Lorentzian shape of the filter passband, the effective bandwidth shrinks rapidly with the number of concatenated filters.

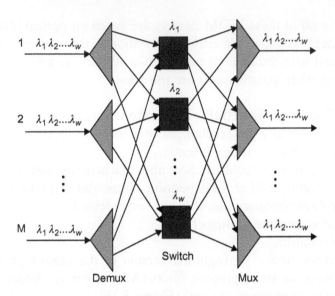

Figure 2.16 Combination of WDM and space switch technologies: W incoming wavelengths, the same for all of the physical input ports, are spatially demultiplexed and distributed to W space switches, each space switch handling one wavelength only. The physical port count for each space switch is M, the number of physical input ports. Total topological port count is M times W. This type of switch is quite versatile; the addition of the wavelength domain can simplify the switch structure. The full exploitation of this technology is reached in the so-called software defined networks, SDNs, an emerging technology aimed at maximizing efficiency in optical fabrics. Some specific types of wavelength selective devices are described here.

REFERENCES

[1] T. Tamir, Integrated optics, in: Topics in Applied Physics, Springer, New York, 1975. ISBN-13: 978-354009

[2] F.A. Blum, K.L. Lawley, W.C. Holton, Naval Res. Rev. 28(1) (1975).

[3] G.P. Agrawal, N.K. Dutta, Long Wavelength Semiconductor Lasers, Van Nostrand Reinhold Electrical/Computer Science and Engineering Series, Springer, New York, 1986.

[4] D.J. Richardson, J. Nilsson, W.A. Clarkson, High power fiber lasers: current status and future perspectives, JOSA B 27 (2010) B63−B92.

[5] A. Yariv, ISBN: 0471971766 Quantum Electronics, second ed., Wiley, New York, NY, 1975.

[6] M.S. Scully, M.S. Zubairy, ISBN: 0521434580 Quantum Optics, Cambridge University Press, UK, 1997.

[7] K. Petermann, ISBN: 90-277-2672-8 Laser Diode Modulation and Noise, Springer, New York, 1988.

[8] E.D. Palik (Ed.), Handbook of Optical Constants of Solids, Academic Press, New York, 1985. ISBN-10: 0125444206.

[9] J.-W. Shi, K.-G. Gan, Y.-J. Chiu, Y.-H. Chen, C.-K. Sun, Y.-J. Yang, J.E. Bowers, Metal−semiconductor−metal traveling-wave photodetectors, IEEE Photon. Technol. Lett. 13 (6) (2001) 623.

[10] B.E.A. Saleh, M.C. Teich, Fundamentals of Photonics, John Wiley & Sons, New York, NY, 1991. ISBN:978-0-471-35832-9, pp. 676–677.

[11] K.-W. Ang, J.W. Ng, G.-Q. Lo, D.-L. Kwong, Impact of field-enhanced band-traps-band tunneling on the dark current generation in germanium p–i–n photodetector, Appl. Phys. Lett. 94 (2009) 223515.

[12] A. Yariv, ISBN: 0030474442 Optical Electronics, fourth ed., Saunders College Publishing, Philadelphia, PA, 1991.

[13] L. Thylen, P. Holmstrom, L. Wosinski, B. Jaskorzynska, M. Naruse, T. Kawazoe, et al., Nanophotonics for low-power switches ISBN:978-0-12-396958-3 in: I.P. Kaminow, T. Li, A. E. Willner (Eds.), Optical Fiber Telecommunications VI, Elsevier Science and Technology Books, Oxford, UK, 2013.

[14] R.A. Spanke, Architectures for large nonblocking optical space switches, IEEE J. Quantum Electron. QE-22 (1986).

[15] M.K. Smit, C. van Dam, PHASAR-based WDM-devices: principles, design and applications, IEEE Sel. Top. Quantum Electron. 2(2) (1996) 236–250.

CHAPTER 3

Silicon and Group IV Photonics

PART ONE: SILICON PHOTONICS ELEMENTS FOR INTEGRATED PHOTONICS

GENERAL PROPERTIES

In the previous chapter, we gave a description of the basic element of integrated photonics, the waveguide, and brief presentations of some of the key corresponding elements in integrated photonics systems. This book is about silicon (and other relevant group IV elements). Silicon made a rather late entrance as a photonics material, though there were some early pioneering papers [1]. The use of silicon as a versatile photonics material is hampered by its indirect bandgap, which

Monolithic Nanoscale Photonics—Electronics Integration in Silicon and Other Group IV Elements.
DOI: http://dx.doi.org/10.1016/B978-0-12-419975-0.00003-9
© 2015 Elsevier Ltd. All rights reserved.

makes it unsuitable for LEDs and lasers, in contrast to the prevalent III−V materials. In addition, it has low absorption only for wavelengths longer than around 1.15 μm. And further more, silicon is centrosymmetric and thus has *no electro-optic (Pockels) effect*, a distinct disadvantage, that has brought about efforts to create such an effect by using strain and other methods of breaking the inversion symmetry. For modulation and other active device functions, the latter can be remedied by using the, albeit weaker, plasma effect [2], where the refractive index change can be controlled by carrier density changes, for example, in p−n junctions, as shown below:

$$\Delta n = - \frac{N_e \lambda_0^2 e^2}{\varepsilon_0 n_s 8 \pi^2 m^* c^2} \equiv -\alpha_n N_e \tag{3.1}$$

where N_e is the free carrier density, λ_0 vacuum wavelength, e electron charge, n_s background index, and m^* the free carrier effective mass.

In silicon, one can write [3] refractive index and absorption changes due to free electrons as

$$\Delta n = - 8.8 \cdot 10^{-22} \, \Delta N_e \tag{3.2}$$

and

$$\Delta \alpha = 8.5 \cdot 10^{-18} \, \Delta N_e \tag{3.3}$$

where the ΔN_e is electron concentration in cm^{-3} and $\Delta \alpha$ attenuation in cm^{-1}.

However, due to its large refractive index (around 3.5 at telecom wavelengths) and comparatively low losses, it makes excellent waveguides for tight light confinement and very small bending radii of waveguides, a prerequisite for dense integration. These features thus enable small devices, an advantage *per se* and further a prerequisite for dense integration and low-power dissipation [4]. Figure 3.1 shows total field width versus core width in microns for a slab waveguide (Figure 2.1(a)), where the core width is infinite at a wavelength of 1.55 μm and for various core and cladding refractive indices. For silicon in air, we have a minimum field width of <500 nm, corresponding to less than ½ of the vacuum wavelength.

Lately, large efforts have been made to find silicon-based group IV alloys that have a direct bandgap, as described in the following section.

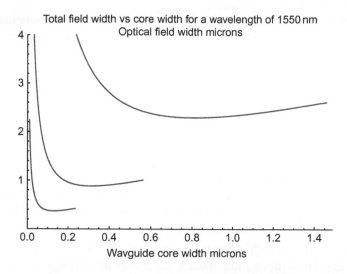

Figure 3.1 Total field width in microns versus core width in microns for a slab waveguide (Figure 2.1) at a wavelength of 1.55 μm and for various core (n_1) and cladding (n_2) refractive indices: from top to bottom $n_1 = 1.5$, $n_2 = 1.4$, as in a glass waveguide, $n_1 = 3.4$, $n_2 = 3.1$ representative for III−V waveguides and $n_1 = 3.5$, $n_2 = 1$, representing silicon in air. As seen, the field width has a minimum for each of the cases; when further reducing the core width, the optical field width increases again. The minimum is smallest for the highest refractive index difference between core and surrounding material. The upper limit of the core width is taken as the core size where the waveguide is no longer single mode but also supports the next higher order mode.

SILICON PHOTONICS ELEMENTS FOR INTEGRATED PHOTONICS: MODULATORS AND WAVELENGTH SELECTIVE DEVICES

Silicon Electro-Optic Modulators

Modulation in silicon devices can be achieved by exploiting the plasma dispersion effect, whereby changing the carrier concentration changes the real and imaginary parts of the index of refraction [1]. In a recent review article, Reed et al. [5] have individuated three possible configurations for electrically driven silicon modulators. In a carrier injection modulator, carriers are injected by forward biasing a p−i−n junction. This type of modulator is limited in speed by the recombination rate of the carriers and has relatively high power consumption. On the positive side, carrier injection allows large changes in the carrier density and therefore high modulation depth. A second method of modulation is carrier depletion. In this case, the modulation in the carrier density is obtained by depleting a p−n junction. Carrier depletion is fast with low power and little current flows in the junction during operation. Because of the need to overlap the p−n junction and the optical field, this method, in general, creates modulators that have large insertion losses and low modulation depth. A third modulation scheme is the charge accumulation. Here an MOS capacitor is used to accumulate charges in the optical waveguide. Like charge depletion, this method leads to fast and low-power modulation. Because of the need to create an oxide barrier in the middle of the modulator waveguide, however, charge accumulation modulators are hard to fabricate.

Two main silicon modulator topologies have been used. Mach−Zehnder interferometers (MZIs) are the simplest and most robust type of modulators and for these reasons have been adapted in many silicon photonic products. An MZI modulator is relatively immune to thermal drifts and fabrication imperfections. MZIs are also relatively large (with typical areas larger or much larger than 1000s of μm^2), have high power consumption, and require complex RF electrode design to achieve high speeds. Green et al. [6] reported a 10 Gb/s injection-based MZI modulator with a total area of $\sim 100 \times 200 \, \mu m^2$ and a power consumption of 5 pJ/bit. Resonant micro-ring modulators have several advantages over MZIs. Because they are resonant devices, micro-rings lend themselves to be used in wavelength multiplexing schemes, in addition they are much smaller than MZIs (with typical areas of 10s of μm^2) and consume much less power. On the negative side, micro-rings are extremely sensitive to thermal shifts and fabrication imperfections. Li et al. [7] have demonstrated a depletion-based ring modulator with a 25 Gb/s modulation rate, a driving voltage of 1 V_{pp}, and 7 fJ/bit power consumption, see Figure 3.2a. Chen et al. [8] demonstrated an injection-based modulator with a 3 Gb/s modulation rate, a driving voltage of 0.5 V, and a power consumption of 86 fJ/bit (Figure 3.2b).

Wavelength Selective Devices in Silicon

Silicon photonics offers superior implementation of the wavelength selective devices as described above due to the high refractive index contrast and plasma effect mediated refractive index control. This facilitates making very small devices and tuning of filter passband frequencies. Disadvantages are the comparatively high sensitivity of the refractive index to temperature changes as well as the increased difficulty in making polarization insensitive devices due to the large index steps normally involved in the waveguides.

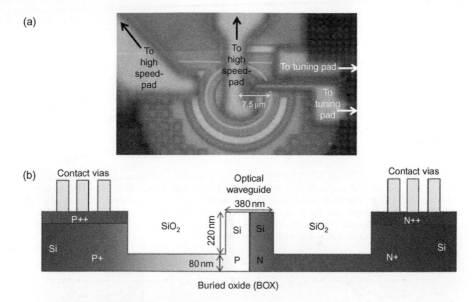

Figure 3.2 Silicon depletion-mode ring modulator with thermal tuning. A 25 Gbps modulation rate, extinction ratio >5 dB, has been demonstrated at a driving voltage of 1 V, corresponding to ∼7 fJ/bit or ∼0.18 mW switching power. The power consumption in the present device is dominated through by the tuning power of up to 66 mW to tune the whole 12.6 nm free spectral range. (a) Photograph of the ring modulator. The upper-right 25% of the ring is made as a Si resistor heater providing wavelength tuning, while 67% is a p−n diode for electro-optic modulation. (b) Cross-sectional diagram of the ring waveguide electro-optic section. The doping profiles of the p−n diode are graded by implant diffusion.

Two types of PICs are representatives of the use of silicon photonics for WDM devices: The **arrayed waveguide grating, AWG** [9], and the **ring resonator**.

The AWG is described in Figures 3.3−3.5.

The Ring Resonator

Figure 3.6 shows the principal layout: light input in port 1 is partly transmitted, partly coupled to the ring at the coupling region between the ring and the straight waveguide. If the ring is not in resonance for the wavelength in question, then light is transmitted to port 2.

The ring resonance condition is

$$N_{\text{eff}} k_0 2\pi r = m 2\pi \tag{3.4}$$

where N_{eff} is the effective index of the ring waveguide, k_0 the wavenumber in vacuum, r the ring radius, and m an integer ≥ 1.

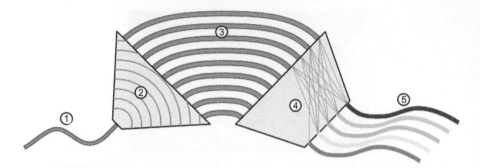

Figure 3.3 Principal physical and functional layout of an AWG: An incident spectrum in waveguide 1 is distributed in a so-called MMI (multimode interference) or free propagation section 2 to an array of waveguides 3. By employing a waveguide array (this has given the device its name) with different optical path lengths, different wavelengths interfere differently in the MMI section 4, resulting in different wavelengths being coupled into different output waveguides. The function is analogous to that of a prism.

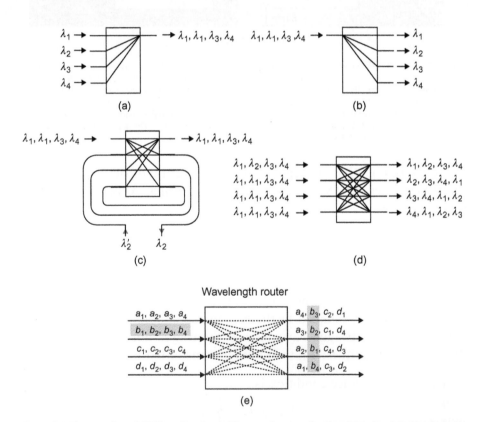

Figure 3.4 The versatility of AWG, making it possible to perform wavelength multiplexing (a), demultiplexing (b), drop/add multiplexing, (c) as well as the cyclic interconnect for wavelength routing (d), better explained in (e). In (e), the four input wavelengths in input 2 from the top (shaded) are distributed over the four outputs (shaded) and corresponding distribution occurs for the same set of four wavelengths at the other input ports. Thus, 4 times 4 inputs are distributed to 4 times 4 outputs and we have a 16 × 16 router.

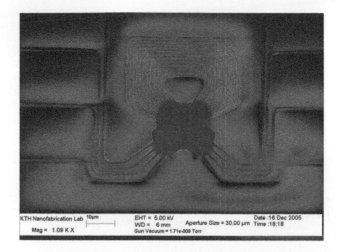

Figure 3.5 A scanning electron microscope (SEM) picture of a 4 × 4 AWG with insertion loss of 6 dB and cross-talk of around −10 dB. This device is based on amorphous silicon and consequently has higher insertion loss than structures made in crystalline silicon. The size is 50 × 40 μm², enabled by the silicon photonics technology with high refractive index [10].

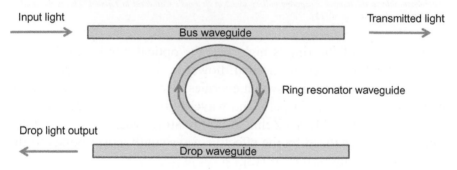

Figure 3.6 Schematic of an optical ring resonator [11]. Without the drop waveguide, the device can be used as modulator as well as a detector, in both cases being wavelength selective. In the former case, the ring resonator is tuned out of phase for maximum transmission and tuned on resonance for decreasing transmission, which is due to absorption of light in the ring; this partial absorption can also be done by a detector. Depending on the resonator Q value, the filter will have different bandwidths.

At resonance, light is coupled into the ring and if a dropping wave-guide is present, eventually coupled to port 4. In the absence of the dropping waveguide and at the so-called critical coupling, light is totally extinguished in waveguide port 2, when the optical field has been built up in the ring. Critical coupling is achieved when the coupling coefficient of the input waveguide to the ring is equal to the losses of the ring. The ring resonator can be converted into a modulator by using the plasma effect described above or using an integrated electro-optic polymer (EOP) to change the refractive index of the ring.

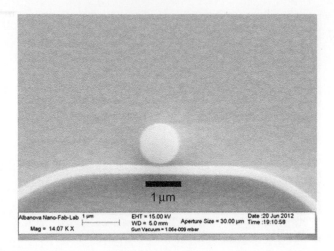

Figure 3.7 Top-view SEM (scanning electron microscope) image of the fabricated microdisk resonator with radius of 525 nm, based on a hybrid plasmonic waveguide structure consisting of a low refractive index material sand-wiched between a gold disk at the top and a silicon disk at the bottom. Note that here one does not use a ring waveguide but rather a disk, where light in the middle low index material is guided inside the perimeter of the disk in a so-called whispering gallery mode, a type of mode peculiar to curved structures for guiding waves. This is a phenomenon akin to the famous whispering gallery sound in St Paul's Cathedral in London. This is the smallest ring resonator reported to date [13].

If the Q value of the ring is high, i.e., the optical bandwidth is small, smaller switch energies are needed, however, at the cost of decreasing bandwidths for decreasing switch energies, demanding high fabrication tolerances to hit the right passband wavelength. Such a trade-off is not so prevalent in a Mach−Zehnder modulator, where the modulator arm length, normalized by the effective wavelength, corresponds to the Q value of the ring resonator [12].

By decreasing the size of the ring resonators, one can achieve more complex structures on smaller footprints reaching to suitable technology with complex switching/routing architectures, very high integration density and modal field confinement below diffraction limit of light, e.g., for very specific applications in optical intercon-nects for inter- and intra-core data communication. Figure 3.7 shows an example of a fabricated hybrid plasmonic disk resonator with a record small radius of only 525 nm [13] and Figure 3.8 shows its experimental spectra (with three different disks−access or, waveguide separations, g). Hybrid plasmonic structure consists of a low refrac-tive index material sandwiched between a gold disk at the top and a silicon disk at the bottom. In such structures, light is highly localized in the low-index dielectric between the metal cap and the high-index

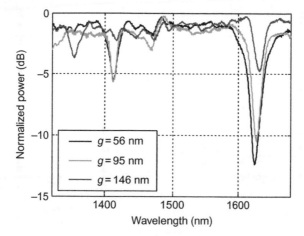

Figure 3.8 Measured transmission spectra of the fabricated hybrid plasmonic microdisks shown in Figure 3.7 with radius of about 525 nm with three different disks—access or bus waveguide separations g (see Figure 3.6). One sharp ring resonance is distinguished, and away from it the maximum transmission is a few dB below input power. The Q value is calculated by dividing the resonance frequency or wavelength by the measured bandwidth 3 dB down from the average maximum transmission. Note that this definition of the Q value only holds for Lorentzian spectral characteristics.

Si. In active devices for switching and routing, one can use EOP instead of the dielectric material and apply switching signal between Au cap and the bottom Si layer.

By including a drop waveguide, light can also be transmitted via the ring to a second waveguide, with the ring in resonance, thus acting as a wavelength selective add drop multiplexer. Adding a wavelength is done in the lower right port, again in resonance with the ring.

Light is coupled between the waveguide and the ring by using the same directional coupling mechanism as shown in Figure 2.11.

PART TWO: BANDGAP ENGINEERING IN GROUP IV MATERIALS FOR PHOTONIC APPLICATION

The bandgap of semiconductors depends on alloying composition, strain, doping level, and media temperature. Figure 3.9(a) and (b) shows a schematic picture of the Si and Ge bandstructure at 300 K. Both materials have a similar valence bandstructure which consists of heavy and light holes and spin-off band. Meanwhile, these materials differ much in conduction band. Ge has a minimum in conduction

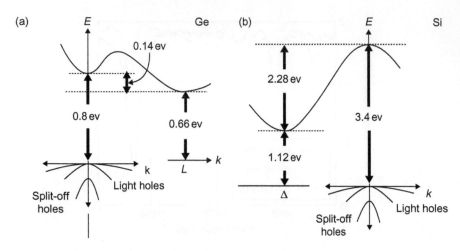

Figure 3.9 Schematic bandstructure of (a) Ge and (b) Si at room temperature.

band at L point creating an indirect bandgap of 0.66 eV, whereas for Si this point lies on Δ-line close to X-point forming 1.12 eV. The heavy and light holes (HHs and LHs, respectively) are defined through their band curvature which determines the effective masses (or inverse of second derivative of $E(K)$).

When Si is alloyed with Ge (in strained or relaxed form), its bandstructure is changed depending on the composition. The SiGe bandstructure is Si-like up to 80% Ge content and it becomes Ge-like for higher Ge content as shown in Figure 3.10. The equations for bandgap for the whole range of Ge content are given as [14]:

$$E_g(x) = 1.155 - 0.43x + 0.0206x^2 \text{ eV} \quad \text{at } 300 \text{ K} \quad x < 0.85 \text{ (Si-like)} \quad (3.5)$$

$$E_g(x) = 2.01 - 1.27x \text{ eV} \quad \text{at } 300 \text{ K} \quad x > 0.20 \text{ (Ge-like)} \quad (3.6)$$

The strain can alter the bandgap structure of Si or Ge when the SiGe layers are grown on Si substrates. The strain consists of two components: hydrostatic and uniaxial components. The hydrostatic part is responsible for the shift of subbands (in conduction or valence band) where uniaxial (or biaxial) strain makes a split of these bands as shown in Figure 3.11.

For example, for compressive-strained SiGe layers, the HH and LH bands are shifted upward and are split where the curvature of these subbands (or effective mass of hole) is changed. The final result is that HH becomes LH-like and LH becomes HH-like.

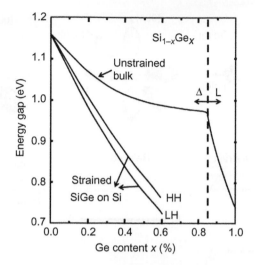

Figure 3.10 The energy gap of SiGe layers versus Ge content [15,16].

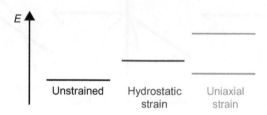

Figure 3.11 The strain components affecting the bandstructure of semiconductor [15,16].

For tensile-strained SiGe layers, the conduction band is modified when the Δ_2 electrons shift downward and are situated below Δ_4 electrons. The effective mass of Δ_2 is lower than Δ_4 electrons and this makes higher mobility (faster carrier transport) for electrons in tensile-strained layers as illustrated in Figure 3.12.

To obtain light emission from Si-based material is a challenge due to the indirect bandgap nature of the material. Both Si and Ge materials have a minimum at Γ-point (the so-called photonic bandgap) which is estimated to be 3.2 and 0.8 eV for Si and Ge, respectively. The indirect bandgap is 1.12 and 0.66 eV which lies at X and L, for Si and Ge, respectively. The energy difference ($\Delta E_{\Gamma-L}$) between the L and Γ for Ge is 140 meV, so it is more probable that Ge exhibits a direct transition compared to Si. Any such transition is still a mixture of both direct and indirect transition. The wavelength of direct transition in Ge material is $\sim$1.55 μm which appears in telecom wavelength region.

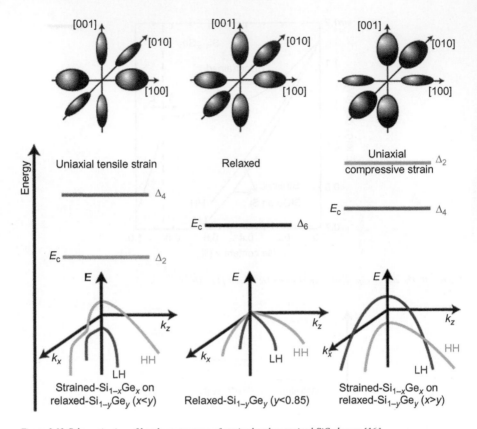

Figure 3.12 Schematic view of bandgap structure of strained and unstrained SiGe layers [16].

Inducing tensile strain in Ge matrix and/or making highly n-type doping improve the luminescence emission efficiency. A pure direct transition may occur when $\Delta E_{\Gamma-L}$ of Ge in conduction band will be as small as possible. By applying strain engineering, the bandgap can be tailored. This can be obtained by alloying Ge with Sn or C atoms. As a response to strain (or alloying), the bandgap narrowing occurs for Ge for both L- and Γ-bands at the same time. Meanwhile, the shift for Γ-band is larger than that of L-band ($\Delta E_{\Gamma-L}$ decreases). In the case of GeSn/Ge system, $\Delta E_{\Gamma-L}$ approaches to zero for Sn contents around 6−8% as shown in Figure 3.13.

Alloying Si with GeSn material results in a ternary system, GeSnSi, where the bandgap is increased from 0.86 upward [18].

$$E_v(Ge_{1-x-y}Si_xSn_y) = E_{v,av}(Ge_{1-x-y}Si_xSn_y) + \frac{\Delta_{so}(Ge_{1-x-y}Si_xSn_y)}{3}$$

$$(3.7)$$

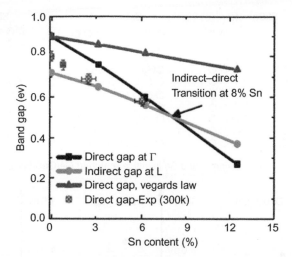

Figure 3.13 The bandgap energy of direct and indirect gaps of GeSn versus Sn content. The indirect-to-direct transition may occur when L and Γ lines intersect with each other [17].

where, $E_{v,av}$ is the average valence band for $Ge_{1-x-y}Si_xSn_y$, which is a linear interpolation and Δ_{so} is spin−orbit splitting.

$$E_{v,av}(Ge_{1-x-y}Si_xSn_y) = -0.48x + 0.69y \tag{3.8}$$

$$\Delta_{so}(Ge_{1-x-y}Si_xSn_y) = 0.295(1 - x - y) + 0.043x + 0.800y$$

For the strain−relaxed and strain-compensated material, the bandgap can also be tailored by alloying. Figure 3.14 shows the bandgap of GeSnSi when the strain is compensated through the fraction rule of Si: Sn of 4:1 in Ge matrix.

Direct-gap values are tunable between 0.8 and 1.4 eV, in GeSiSn alloys, epitaxially grown on Ge-buffered Si, as a function of the combined Si + Sn fraction X [18]:

$$E_0(X) = E_{0Ge} + AX + BX^2, \quad \text{where } A = 1.70 \pm 0.42 \text{ and } B = -1.62 \pm 0.96 \tag{3.9}$$

$Ge_{1-x-y}Si_xSn_y$ materials demonstrate high thermal stability far behind GeSn layers which show both phase separation and Sn precipitates for temperatures above 500°C rapid thermal annealing (RTA) [19,20].

The relative thermal stability of $Ge_{1-x}Sn_x$ growth should be below 350°C; meanwhile, for $Ge_{1-x-y}Si_xSn_y$, the growth is limited to 380°C and the alloys are thermally stable up to 850°C RTA [21].

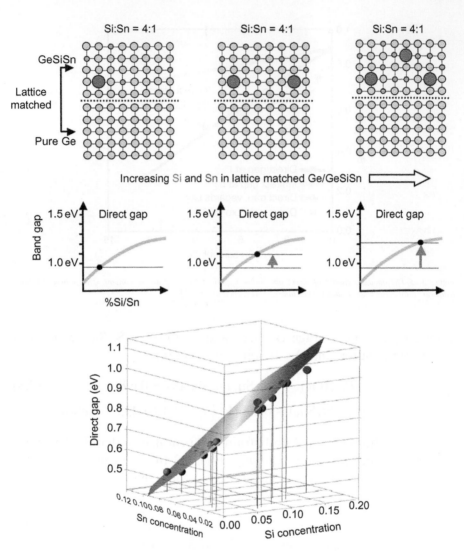

Figure 3.14 Schematic of Si:Sn fraction in strain compensated GeSnSi layers and estimated energy bandgap [18].

This remarkable improvement in thermal stability can be explained by the larger entropy of mixing for GeSnSi alloys [21].

PART THREE: GROUP IV PHOTODETECTORS

Historically, the first Si-like photodetectors were manufactured in 1980s with a PIN profile where $Si_{1-x}Ge_x$ ($0.40 \leq x \leq 0.60$)/Si superlattice (SL)

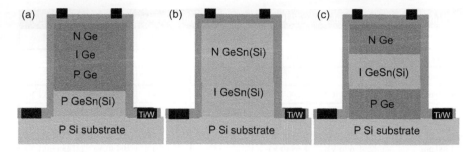

Figure 3.15 PIN structures in the form of (a) GeSn(Si)-VS for tensile strained Ge, (b) strain−relaxed GeSn (Si), and (c) compressive strained GeSn(Si) on Ge-VS.

demonstrated responsivity of 423 mV/A at 1300 nm. High Ge content was desired in these detectors' structure in order to obtain absorption in telecom wavelengths. High-quality SiGe (6 nm)/Si (29 nm) with 20 periods were grown with low defect density. Although these detectors were carefully manufactured, their absorption coefficient was decreased significantly for larger wavelengths due to the quantum confinement effect resulted from thin SiGe layers in SLs [22].

A better solution was proposed by integrating SiGeC in the intrinsic part of PIN detectors where the strain in the SiGe was compensated by carbon. In these detectors, 80 nm SiGeC layer was grown with high epitaxial quality and the absorption coefficient was estimated to $1.0 \times 10^{-3} \, \mu m^{-1}$ at 1.55 μm and $3.4 \times 10^{-2} \, \mu m^{-1}$ at 1300 nm [23].

To further improve the absorption coefficient and responsivity of the detectors, Ge and Ge-like materials are excellent candidates for telecom wavelengths and near-infrared (NIR) region. This means that layers, e.g., $Ge_{1-x}Si_x$ ($x < 0.20$), $Ge_{1-y}Sn_y$ (strained or relaxed), and $Ge_{1-x-y}Si_xSn_y$ alloys, and tensile-strained Ge are the most appropriate materials to be grown on (virtual substrate) Ge. There are two ways to design a photodetector made of group IV materials. To integrate Sn alloys in the active region of PIN (e.g., as intrinsic) or as a virtual substrate (VS) for Ge as active material are novel designs for advanced detector structures. Figure 3.15 shows three designs for GeSn(Si) and Ge materials in photodetectors.

The detectors consist of mesa structures with different sizes which are passivated by SiO_2. A top and bottom contact with integrated NiGe prior to metallization (Ti/W or Cr/Au) is formed. In some cases, a thin Si cap layer is grown to ensure a better contact and make less reflectance.

Table 3.1 The Measured Responsivity of GeSn Photodetectors with Different Design						
R (mA/W)	Ref. [24]	Ref. [25]	Ref. [26]	Ref. [27]	Ref. [28]	Ref. [29]
Layer thickness (nm)	300	430	820	750	300	300
Sn content (%)	0.5	2	3	3.6	3.85	4
1550 nm	102 (0 V)	200 (0 V)	230 (0.1 V)	275 (0 V)	272 (0 V)	178 (−0.1 V)
1600 nm	45 (0 V)		154 (0.1 V)			
1640 nm	18 (0 V)	167 (0 V)	120 (0.1 V)	255 (0 V)	166 (0 V)	162 (−0.1 V)
1700 nm	6 (0 V)	69 (0 V)		28 (0 V)	104 (0 V)	145 (−0.1 V)
1800 nm		15 (0 V)		325 (0 V)	8 (0 V)	96 (−0.1 V)
The applied voltages for each case are shown in the parenthesis.						

Table 3.1 demonstrates the characteristic data from PIN detectors with GeSn(Si) in intrinsic part (Figure 3.15(c)). The operating wavelengths of these detectors are ranging from 1.55 to 1.80 μm depending on Sn content and the strain in the active region.

The indirect-to-direct transition may occur in GeSn layers for Sn content above 6% (the energy of Γ-band lies below L-band). This means that all the detectors in Table 3.1 will have a transition from both Γ- and L-bands. New gas precursors provide possibility to CVD technique to further decrease the growth temperature and impede the Sn segregation. For 1600−2000 nm wavelengths, the Ge detectors are considered as competitors with AlGaAs-based detectors. However, there are some differences between Ge- and AlGaAs-based detectors. As it was discussed in Chapter 2, the shunt resistance for the detector is important and it should be kept as large as possible in order to decrease the Johnson noise in the detector. The InGaAs detectors exhibit shunt resistance in range of MΩ, meanwhile for Ge detector this is in kΩ level. Therefore, Ge photodetectors demonstrate a remarkably higher thermal noise compared to InGaAs detectors.

The dark current of Ge-based detectors is usually significantly higher than InGaAs detectors. It is important to emphasize here that if the active area of germanium detector is large, the dark current will be large (higher shot noise as well) and shunt resistance is low (higher thermal noise). In this case, the detectors have to be manufactured with small size to impede the dark current.

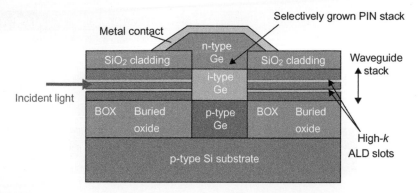

Figure 3.16 A Ge structure with PIN profile is grown selectively for photodiodes where the waveguide is integrated [30].

INTEGRATION OF PHOTODIODES WITH WAVEGUIDE OR MOSFETs

The light wave guiding is the mainstream for the future communication technology where the data is transmitted core-to-core in computers or rack-to-rack between communication servers. The wavelength range of $1.33-1.55\,\mu m$ is the practical interval for data communication for today's fiber-optics. Si photonics proposes a cost-effective technology compared to III–V and will expand in large range of applications in the near future.

Si photonic materials are grown primarily on Si wafers and are processed to form a mesa structure. The selective epitaxy growth provides the possibility to deposit a structure on a patterned Si wafer where mesa can be grown after fabrication of the waveguides. The grown mesa can be detectors, modulators, or lasers. For example, in the case of PIN detector, the light may illuminate the intrinsic layer through a Si waveguide as shown in Figure 3.16. A drawback with selective epitaxy is the pattern dependency leading to nonuniform structure profile. The pattern layout (density and opening sizes) has to be modified to create a uniform consumption of reactant gases leading to uniform deposition over the entire wafer as described in Chapter 1.

Low loss amorphous Si waveguides are suitable to integrate the transistor-photonics with other optical components on the same chip for the modern CPU.

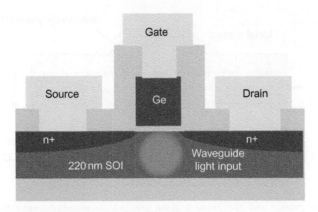

Figure 3.17 Schematic cross-section of the photoMOSFET [34].

PHOTOMOSFETs

The design for advanced on- and off-chip data communications, the electrical wires will be entirely replaced by photonic components and optical interconnects. Being used the connecting metal lines show a capacitance of ~ 0.2 fF/μm which is in practice too high for future interconnections. Today's goal is the energy per bit receivers should be decreased to a level of 10 fJ/bit [31].

This requires photodiodes and other devices with low capacitance and all components should be densely integrated. Therefore, many research activities [32,33] have focused to integrate the photodiodes monolithically with complementary metal-oxide-semiconductor (CMOS) receiver circuitry [34].

One way to pass through these problems is to integrate MOSFETs and photonic components together. Recent research shows that this alternative is possible and photodetector and transistor are manufactured in a single body, the so-called photoMOSFET (Figure 3.17).

In this device, the photodiode is reverse biased. When light is illuminated to the germanium gate [21] a photovoltage is generated across the germanium. This photovoltage increases the total applied voltage to the gate contact. This voltage is obtained from equation (3.10) where η is quantum efficiency, P_{inc} denotes the incident optical power, I_s stands for the leakage current of diode and ν is the frequency [35].

$$V_{\text{photo}} = \frac{nkT}{q} \ln \left(\frac{P_{\text{inc}}}{I_s} \frac{\eta q}{h\upsilon} \right) \qquad (3.10)$$

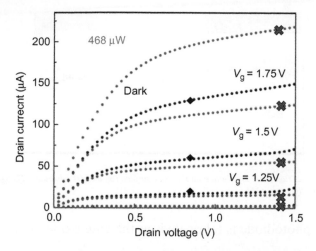

Figure 3.18 Drain current versus gate voltage both in the dark (◇) and with light illumination of 468 μW(×) [34].

This condition will have an impact on ultimate device performance. From this we can then calculate the photocurrent in the drain current, I_d, when the device is in saturation using the simple square law current equation.

When the device is in saturation mode, the photocurrent is calculated from

$$I_d = \frac{1}{2}\mu_n C_{ox}\frac{W}{L}(V_g + V_{photo} - V_T)^2 \tag{3.11}$$

where the transistor's gate length and width are L and W, respectively, electron mobility is μ_n, gate oxide capacitance, C_{ox}, the threshold voltage, V_T, and an applied gate bias is V_g.

The device demonstrates a photocurrent of 51 μA when a bias current of 150 μA is applied and a light beam of 468 μW is illuminated (as shown in Figure 3.18).

Equation (3.11) shows that the drive current is influenced by induced photovoltage which is also sensitive to both the quantum efficiency of the absorption and the leakage current of internal diode. The latter parameter is dependent on the defect density in the photodiode. The quantum efficiency of the absorption depends on the level of n-doping in Ge and the thickness of the gate oxide.

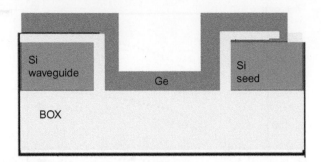

Figure 3.19 A cross-section of Ge layer grown on the oxide surface (on an SOI substrate). The Ge layer has been laterally and vertically grown [35].

The Ge photodiode is fabricated on the gate oxide in gate region of transistor.

This Ge layer is polycrystalline if it is directly grown on the oxide surface. However, the layer quality can be improved by a new method, the so-called rapid melt growth (RMG) [35]. In this method, the Ge layer is grown over oxide surface when a Si seed region is patterned and prepared for selective epitaxy (Figure 3.19). The Ge layer is grown from the seed layer over the oxide region. A spike annealing at 1000°C for 1 s is necessary to improve the quality of the epitaxial layer.

Integration of photodiodes on the downscaled transistors will improve the device performance in terms of speed and sensitivity for future low-energy optical communication.

GROUP IV-BASED LASERS

A lasing action from group IV materials with indirect bandgap property can occur through transitions in the valence or conduction subband minima where the energy difference between the subbands can be tailored by inducing strain in the structure. These intersubband transitions (ISTs) are mixtures of photons and phonons as shown in Figure 3.20.

Early works on $Si_{1-x}Ge_x/Si$ structure for IST design showed promising structures for quantum cascade laser (QCL) [36,37].

These devices function in regions of NIR (1−2 μm) or middle infrared (2−20 μm) depending on Ge content in $Si_{1-x}Ge_x/Si$ (for $0.25 \leq x \leq 0.50$)

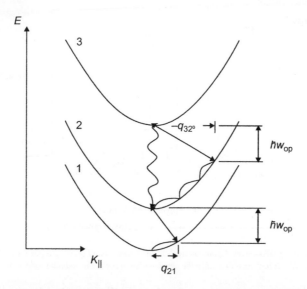

Figure 3.20 Photon and phonon scattering in three subbands in a group IV material with indirect bandgap when laser pumped the subband 3. q32 and q21 stands for phonon momenta and ħω presents the phonon energy.

are available wavelengths in the range of $5-20$ μm, where ΔE_v lies in the range of $205-420$ meV.

The envelope functions can be obtained by solving the Schrodinger equations for quantum wells. The QCL structure consists of an injector and an active part. By applying a voltage, the carriers are injected into active part.

In QCL, the envelope functions are engineered by choosing materials which create barrier heights and quantum wells while the barrier widths adjust the subband energies and spatial overlap of envelope functions into the neighboring subbands. In this case, the carriers are confined in the z-direction and are unconfined in the x- and y-direction [38] (Figure 3.21).

One problem with SiGe/Ge is that the band offset is essentially in the valence band and holes are the carriers for transition. Since the holes have larger effective mass than electrons, therefore, it is better to design structures for transitions with conduction band offset.

The new group IV alloying of Sn−Ge−Si gives a hope for a new generation of lasers. The high mobility MOSFET with GeSn in channel region has been already demonstrated as well as different GeSn

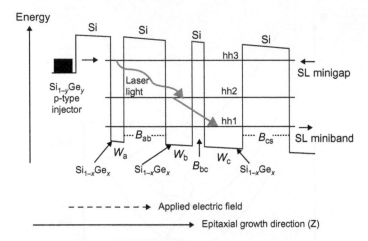

Figure 3.21 Schematic of three-coupled quantum wells in the valence band of $Si_{1-x}Ge_x/Si$ QCL. The width of barriers $(W_a, W_b,$ and $W_c)$ is increased to modify the envelope functions in the quantum wells [38].

PIN photodetectors; however, an electrically injected laser diode which is the key device for monolithic integration of Si-based photonic components with electronic ones on the chip is still missing.

The significant advantage of Sn alloys is the possibility to obtain an indirect-to-direct bandgap transition when 6–8% Sn is incorporated in GeSn matrix. These GeSn layers have narrower bandgap than Ge layer and they can be used for radiative band-to-band emission in the IR region. Optically pumped Ge-on-Si lasers for wavelength of 1.6 μm has already been shown [39]. Recent simulations propose suitable alloy of Ge−Sn−Si materials for electrically pumped lasers as light sources:

1. Lattice-matched $Ge/Ge_{1-x-y}Si_xSn_y$ multilayers for ISTs in a QCL structure which functions at the L-valley.
2. Lattice-matched GeSn/GeSiSn structures in the form of a multiple quantum well (MQW) of functions on band-to-band (or interband) transitions in type I active which functions at Γ-valley (Figure 3.22).

Ternary $Ge_{1-x-y}Sn_ySi_x$ layers offer the possibility to modify independently the bandstructure and strain by adjusting the Si and Sn contents in the alloys (see Figure 3.14) [19]. The calculated energy bandgap of GeSnSi materials shows that a lattice-matched $Ge/Ge_{0.76}Si_{0.19}Sn_{0.05}$ structure is an appropriate structure for QCL [42]. The energy difference between the L- and Γ- or X-valleys is ~150 meV and both these valleys are located above L-valley. Therefore, in a multilayer structure of $Ge/Ge_{0.76}Si_{0.19}Sn_{0.05}$, Ge layers are quantum wells and

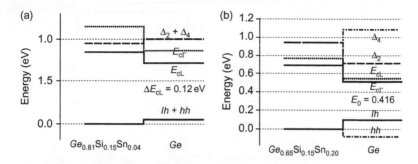

Figure 3.22 Illustration of laser design: (a) IST at L-valley [40] and (b) interband transition at Γ-valley [41].

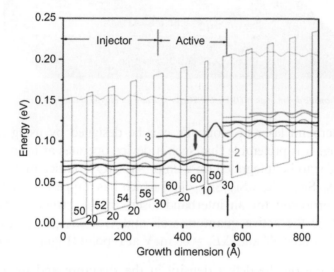

Figure 3.23 Schematic structure of QCL consists of a MQW structure of Ge/Ge$_{0.76}$Si$_{0.19}$Sn$_{0.05}$, where Ge layers are active quantum wells of the laser [43].

Ge$_{0.76}$Si$_{0.19}$Sn$_{0.05}$ layers are barriers. Figure 3.23 shows a schematic of L-valley conduction band of MQWs of Ge/Ge$_{0.76}$Si$_{0.19}$Sn$_{0.05}$ in a QCL structure. The MQW–QCL structure can be grown on Si if a thin Ge seed is grown on Si first.

It is important to mention here that the conventional waveguides are not suitable for QCL operating far infrared region and a double-Au-plasmon waveguide can be applied for these structures [40,44]. In this case, an MQW–QCL can be sandwiched between two Au layers and the detector body needs a mirror with a cavity length of 1 mm. A threshold current density of 550 A/cm^2 at 300 K is determined for such laser design.

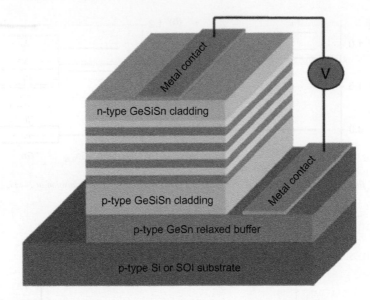

Figure 3.24 Schematic design of an MQW GeSnSi/GeSn structure for interband laser [44].

An interband laser in group IV can be designed by using advanced engineering. It is preferable that such a device operates at room temperature in order to be integrated on a chip. An MQW lattice-matched structure of $Ge_{0.9}Sn_{0.1}/Ge_{0.75}Si_{0.1}Sn_{0.15}$ [44] which has a type I band alignment could be proposed for an interband laser. In this structure, $Ge_{0.9}Sn_{0.1}$ layers act as the active quantum well and $Ge_{0.75}Si_{0.1}Sn_{0.15}$ as barriers where $\Delta E_c = 88$ meV and $\Delta E_v = 68$ meV at Γ point (Figure 3.24).

Depending on the defect density in the structure and the choice of materials in the active region, the electron−hole recombination can result in phonon (Auger or irradiative) or photon generation.

The lower DOS in the MQWs decreases the Auger recombination which leads to the condition where the carrier lifetime is ruled by the radiative recombination process.

In order to have a decent optical confinement factor, the number of periods of QWs is important. 35 periods of $Ge_{0.9}Sn_{0.1}/Ge_{0.75}Si_{0.1}Sn_{0.15}$ will result in 90% optical confinement.

To characterize the laser performance, modal gain is usually discussed. This parameter is a product of the optical gain and mode confinement factor. The optical gain is estimated by the energy difference between the lowest ground state for the electrons in the subband and the uppermost ground state for HH subband.

For a laser, the modal gain has to be high enough to compensate for the different losses in the device, e.g., the carrier absorption and imperfect mirror reflectivity.

PART FOUR: GRAPHENE, NEW PHOTONIC MATERIAL

Among graphene's mechanical, optical, and electrical characteristics, the most promising is its optical property. Graphene material has been already integrated in devices, e.g., modulators [45,46], plasmonic structures [47], and photodetectors [48,49] covering all optical communication bands.

Graphene has distinctive optical properties. Despite its ultimate thinness, graphene absorbs about 2.3% of white light as a result of its unique electronic structure [50]. It absorbs a large range of wavelengths through interband, intraband, and collective plasmon excitations. The exceptional advantage of graphene is in the tunability of absorption by adjusting the charge density.

For the NIR and visible wavelengths, the photon energies lie in the linear part of the graphene dispersion relation where interband excitations take place. In this range, the optical conductivity is given by a universal value: $\sigma_{uni} = \pi e^2/2h$. In higher photon energies, corresponding gradually to nonlinear parts of dispersion relation, the optical conductivity increases smoothly until it peaks pronouncedly at $E_{exp} = 4.62$ eV (in UV range) which can be described by interband transitions in graphene from the bonding to the antibonding π states near the M point of the Brillouin zone [51] (Figure 3.25).

For far IR and terahertz range, graphene's intraband excitations dominantly take place, as in metals, principally. However, in graphene, these excitations can be tuned by electrostatic doping. The optical conductivity in this range is given by

$$\alpha(\omega) = \frac{i2E_F\sigma_{uni}/\hbar}{\pi(\omega + i\Gamma)} \tag{3.12}$$

where Γ^{-1} is damping rate [52] and the Fermi energy level (E_F) can be adjusted by electrostatic doping, e.g., by a gate voltage.

The absorption of light in lateral direction when a graphene sheet is integrated on a waveguide is determined by the length of graphene

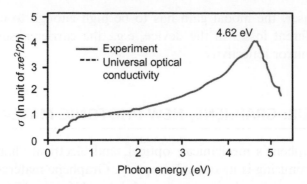

Figure 3.25 *Experimental optical conductivity (solid line) and the universal optical conductivity (dashed line) of monolayer graphene. Note the deviation of the optical conductivity from the universal value at low energies is attributed to spontaneous doping [51].*

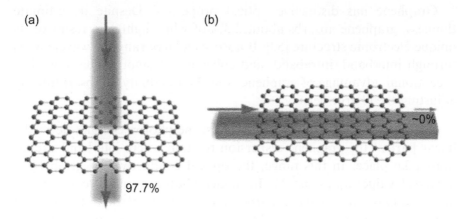

Figure 3.26 *(a) Normal incident light on a graphene sheet, short interaction length with absorption coefficient of 2.3%. (b) Graphene on a waveguide; the light–graphene interaction is tailored by length of the graphene sheet [53].*

(or waveguide) sheet and a complete absorption can be obtained if the length of device is long enough (Figure 3.26).

The normal incident light absorption in graphene occurs in intra- and interband transitions as shown in Figure 3.27. The optical conductance curve may differ from the ideal universal value of 2.3% depending on the beneath layer. This deviation related to phonons where interband absorption is allowed in the presence of phonons [54].

The absorption can be tailored for a particular wavelength by applying a voltage to adjust the carrier concentration and type in the graphene material which has a low DOS.

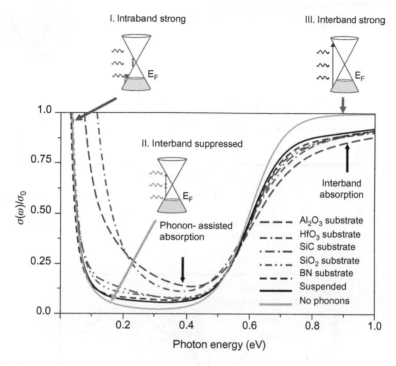

Figure 3.27 Calculated frequency dependence of the optical conductivity of graphene ideally or on several different substrates with impurity concentration of 5×10^{11} cm^{-2} and chemical potential of 0.3 eV at room temperature [54].

It has been already demonstrated that a graphene modulator processed on a Si waveguide can function up to 1 GHz. In this device, the contact is formed through highly doped Si in an extension region. A more convenient way to deal with graphene modulator is illustrated in Figure 3.28 where two graphene sheets separated by an oxide layer are acting in modulation. The first sheet layer is intended for absorption, but the second one works as a gate. Such a design is realized if the thickness of the second layer is thicker than the first one. SiO_2 has been used in this device between the graphene sheets.

Silicon can be used as a waveguide where it is transparent for 1.1 μm $< \lambda <$ 3.6 μm, for larger wavelengths the absorption loss is increased substantially. For the wavelengths above 3.6 μm, a suspended Si membrane or Ge (on Si) can be used as waveguide [14]. In this modulator operates, the 136 and 325 nm for $\lambda = 1.55$ μm and 3.5 μm, respectively.

The modulator may operate when the bottom graphene is ground and the top one is raised with a gate voltage, V_g [46] (Figure 3.29).

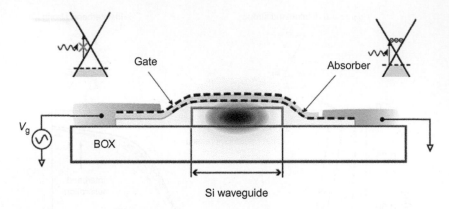

Figure 3.28 A dual-layer graphene modulator integrated on a waveguide. Applying a gate voltage in this device may change the sheet carrier densities in the bottom graphene sheet [46].

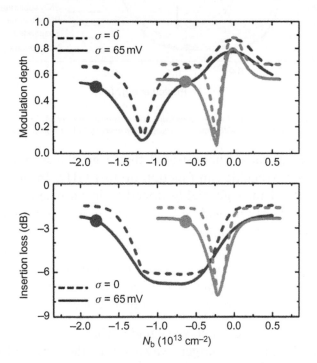

Figure 3.29 A graphene-based modulator operating at $\lambda < 1.55\ \mu m$ (curves in left) and $\lambda < 3.5\ \mu m$ (curves in right) when voltages 0 and a 65 mV (fluctuated potential) are applied to the gate, then (a) modulation depth and (b) insertion loss vary versus the electron concentration, n_b [46].

PHOTODETECTORS

Graphene as material for a photodetector was first shown when a detectable current was generated by illuminating the areas close to electrical contacts of a graphene transistor [55,56]. Graphene has a low

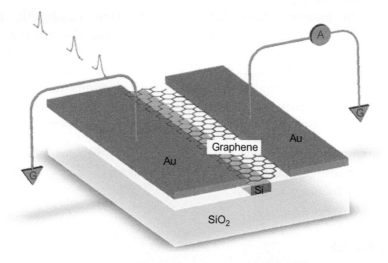

Figure 3.30 Process steps: SiO₂ is deposited planarized to obtain a flat surface for graphene sheet deposition. Another SiO₂ layer with thickness of 10 nm is deposited as an isolation layer for graphene sheet and Si waveguide. An optical absorption may occur in interaction with light in waveguide and generate photocurrent. Responsivity: 15 mA/W [49].

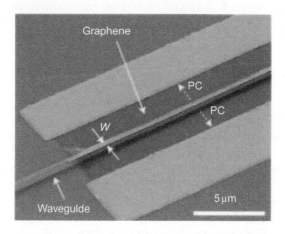

Figure 3.31 Process steps: (1) Si waveguide is formed by etching and then is passivated, (2) deposition graphene and patterning (3) metallization. Both the metal contacts are grounded creating a GND-S-GND configuration (GND: ground and S: signal electrode) which provides a doubling of photocurrent (PC). The central electrode is Ti/Au and graphene generates a PC. Responsivity: 30–50 mA/W [48].

optical absorption but by integrating a graphene photodetector on a bus waveguide formed on an SOI substrate, it is mostly probable to increase the graphene absorption and the resulted photodetection. Figures 3.30 and 3.31 illustrate the two graphene photodetectors.

Although graphene-based detectors have high bandwidth, their responsivity is low (<1 mA/W) due to the low optical absorption in graphene in vertical illumination. The plasmon frequency can be tuned from NIR to terahertz range. As well as chemical and electrostatic doping, graphene engineering by making ribbons of graphene can also be used to tune the plasmon frequency. The frequency is reversely proportional to the ribbon width ($\omega_{pl} \propto w^{-1/2}$, where w is the ribbon width) [57].

Photocurrent in graphene can be generated by various mechanisms including photovoltaic effect, photothermoelectric effect, bolometric effect, and phonon drag effect.

REFERENCES

[1] R. Soref, B. Bennett, Electrooptical effects in silicon, IEEE J. Quantum Electron. 23 (1987) 123–129.

[2] Integrated optics, in: T. Tamir, Topics in Applied Physics, Springer, New York, 1975 ISBN-13: 978-3540096733 D.K. de Vries, Investigation of gross die per wafer formulas, IEEE Trans. Semicond. Manuf. (2005) 136–139. Available from: http://dx.doi.org/doi:10.1109/TSM.2004.836656.

[3] O. Limon, Z. Zalevsky, L. Businarob, Metal-oxide semiconductor, field effect transistor based microscale electro-optical multimode interference modulator on a silicon chip, J. Nanophotonics 1 (2007) 011660.

[4] L. Thylen, P. Holmstrom, L. Wosinski, B. Jaskorzynska, M. Naruse, T. Kawazoe, et al., Nanophotonics for low-power switches, in: I.P. Kaminow, T. Li, A.E. Willner (Eds.), Optical Fiber Telecommunications VI, Elsevier Science and Technology Books, Oxford, UK, 2013.

[5] G.T. Reed, G. Mashanovich, F.Y. Gardes, D.J. Thompson, Silicon optical modulators, Nat. Photon. 4 (2010) 518–526.

[6] W.M. Green, M.J. Rooks, L. Sekaric, Y.A. Vlasov, Ultra-compact, low RF power, 10 Gb/s silicon Mach–Zehnder modulator, Opt. Express 15 (2007) 17106–17113.

[7] G. Li, X. Zheng, J. Yao, H. Thacker, I. Shubin, Y. Luo, et al., 25 Gb/s 1V-driving CMOS ring modulator with integrated thermal tuning, Opt. Express 19 (2011) 20435–20443.

[8] L. Chen, K. Preston, S. Manipatruni, M. Lipson, Integrated GHz silicon photonic interconnect with micrometer-scale modulators and detectors, Opt. Express 17 (2009) 15248–15256.

[9] X.J.M. Leijtens, M.K. Smit, Arrayed waveguide gratings, in: Wavelength Filters in Fiber Optics, 2006, doi:10.1007/3-540-31770-8_5.

[10] D. Dai, L. Liu, L. Wosinski, S. He, Design and fabrication of ultra-small overlapped AWG demultiplexer based on α-Si nanowire waveguides, Electron. Lett. 42 (2006) 400–402.

[11] Wim Bogaerts, et al., Silicon microring resonators, Laser Photonics Rev. 6 (2012) 47–73.

[12] L. Thylen, A comparison of optically and electronically controlled optical switches, Appl. Phys. A A113 (2013). Available from: http://dx.doi.org/doi:10.1007/s00339-013-7914-x.

[13] F. Lou, L. Thylen, L. Wosinski, Hybrid plasmonic microdisk resonators for interconnect applications, in: Optics + Optoelectronics International Symposium, 15–18 April 2013, Prague, Czech Republic, SPIE Proc., 2013.

[14] J. Weber, M.I. Alonso, Near-band-gap photoluminescence of Si−Ge alloys, Phys. Rev. B 40 (1989) 5683.

[15] E. Kasper, Properties of Strained and Relaxed Silicon Germanium ISBN: 0852968264 INSPEC, London, United Kingdom, 1995.

[16] D.J. Paul, Si/SiGe heterostructures: from material and physics to devices and circuits, Semicond. Sci. Technol. 19 (2004) R75.

[17] S. Gupta, R. Chen, B. Magyari-Kope, H. Lin, B. Yang, A. Nainani, et al., GeSn technology: extending the Ge electronics roadmap, in: IEEE International Electron Devices Meeting, IEDM 16.6.1, 2011, p. 398.

[18] V.R. D'Costa, Y.-Y. Fang, J. Tolle, A.V.G. Chizmeshya, J. Kouvetakis, J. Menéndez, Tunable Optical Gap at a Fixed Lattice Constant in Group-IV Semiconductor Alloys, Phys. Rev. Lett. 102 (2009) 107403.

[19] B. Vincent, F. Gencarelli, H. Bender, C. Merckling, B. Douhard, D.H. Petersen, et al., Undoped and in-situ B doped GeSn epitaxial growth on Ge by atmospheric pressure-chemical vapor deposition, Appl. Phys. Lett. 99 (2011) 152103.

[20] Y. Shimura, N. Tsutsui, O. Nakatsuka, A. Sakai, S. Zaima, Control of Sn Precipitation and Strain Relaxation in Compositionally Step-Graded $Ge_{1-x}Sn_x$ Buffer Layers for Tensile-Strained Ge Layers, Jpn. J. Appl. Phys. 48(Part 1) (2009) 04C130.

[21] C. Xu, L. Jiang, J. Kouvetakis, J. Menendez, Optical properties of $Ge_{1-x-y}Si_xSn_y$ alloys with $y > x$: direct bandgaps beyond 1550 nm, Appl. Phys. Lett. 103 (2013) 072111.

[22] H. Temkin, T.P. Pearsall, J.C. Bean, R.A. Logan, S. Luryi, $Ge_x Si1_x$ strained layer superlattice waveguide photodetectors operating near 1.3 μm, Appl. Phys. Lett. 48 (1986) 963. Available from: http://dx.doi.org/doi:10.1063/1.96624.

[23] F.Y. Huang, S.G. Thomas, M. Chu, K.L. Wang, Epitaxial SiGeC/Si photodetector with response in the 1.30−1.55 μm wavelength range, IEDM96, 1996, pp. 665−668.

[24] J. Werner, M. Oehme, M. Schmid, M. Kaschel, A. Schirmer, E. Kasper, et al., Germanium−tin p−i−n photodetectors integrated on silicon grown by molecular beam epitaxy, Appl. Phys. Lett. 98 (2011) 061108.

[25] R.T. Beeler, J. Gallagher, C. Xu, L.Y. Jiang, C.L. Senaratne, D.J. Smith, et al., Bandgap-engineered group-IV optoelectronic semiconductors, photodiodes and prototype photovoltaic devices, ECS J. Solid State Sci. Technol. 2 (2013) Q172.

[26] S. Su, B. Cheng, C. Xue, W. Wang, Q. Cao, H. Xue, et al., GeSn p−i−n photodetector for all telecommunication bands detection, Opt. Express 19 (2011) 6400.

[27] D. Zhang, C. Xue, B. Cheng, S. Su, Z. Liu, X. Zhang, et al., High-responsivity GeSn short-wave infrared p−i−n photodetectors, Appl. Phys. Lett. 102 (2013) 141111.

[28] H.H. Tseng, H. Li, V. Mashanov, Y.J. Yang, H.H. Cheng, G.E. Chang, et al., GeSn-based p−i−n photodiodes with strained active layer on a Si wafer, Appl. Phys. Lett. 103 (2013) 231907.

[29] M. Oehme, M. Schmid, M. Kaschel, M. Gollhofer, D. Widmann, E. Kasper, et al., GeSn p−i−n detectors integrated on Si with up to 4% Sn, Appl. Phys. Lett. 101 (2012) 141110.

[30] M.M. Naiini, H.H. Radamson, G. Malm, M. Östling, Embedded PIN Germanium Photodetectors in High-k ALD Slot Waveguides, 2014 15th Int. Conf. Ultimate Integration on Silicon (ULIS) (2014) 45.

[31] D.A.B. Miller, Device requirements for optical interconnects to silicon chips, Proc. IEEE 97 (2009) 1166−1185.

[32] H. Pan, S. Assefa, W.M.J. Green, D.M. Kuchta, C.L. Schow, A.V. Rylyakov, et al., High-speed receiver based on waveguide germanium photodetector wirebonded to 90 nm SOI CMOS amplifier, Opt. Express 20 (2012) 18145.

[33] S. Assefa, H. Pan, S. Shank, W.M.J. Green, A. Rylyakov, C. Schow, et al., Monolithically integrated silicon nanophotonics receiver in 90 nm CMOS technology node, presented at the Opt. Fiber Commun. Conf. Expo. Nat. Fiber Opt. Eng. Conf., Anaheim, CA, 2013, pp. 1–3.

[34] R.W. Going, J. Loo, T.-J. King Liu, M.C. Wu, Germanium gate photo-MOSFET integrated to silicon photonics, IEEE J. Sel. Top. Quantum Electron. 20 (2014) 8201607.

[35] Y. Liu, M. Deal, J. Plummer, Rapid melt growth of germanium crystals with self-aligned microcrucibles on Si substrates, J. Electrochem. Soc. 152 (2005) G688–G693.

[36] I. Bormann, K. Brunner, S. Hackenbuchner, G. Abstraiter, S. Schmult, W. Wegscheider, Midinfrared intersubband electroluminescence of Si/SiGe quantum cascade structures, Appl. Phys. Lett. 80 (2002) 2260–2262.

[37] S.A. Lynch, R. Bates, D.J. Paul, D.J. Norris, A.G. Cullis, Z. Ikonic, et al., Intersubband electroluminescence from Si/SiGe cascade emitters at terahertz frequencies, Appl. Phys. Lett. 81 (2002) 1543–1545.

[38] R. Soref, L. Friedman, G. Sun, Silicon intersubband lasers, Superlattices Microstructures. 23 (1998) 427–439.

[39] J. Liu, X. Sun, R. Camacho-Aguilera, L.C. Kimerling, J. Michel, Ge-on-Si laser operating at room temperature, Opt. Lett. 35 (2010) 679–681.

[40] G. Sun, H.H. Cheng, J. Menendez, J.B. Khurgin, R.A. Soref, Strain-free Ge/GeSiSn quantum cascade lasers based on L-valley intersubband transitions, Appl. Phys. Lett. 90 (2007) 251105.

[41] S.-W. Chang, S.L. Chuang, Theory of optical gain of Ge-$Si_xGe_ySn_{1-x-y}$ quantum-well lasers, IEEE J. Quantum Electron. 43 (2007) 249–256.

[42] G. Sun, Towards Si-based electrically injected group-IV lasers, Opt. Quantum Electron. 44 (2012) 563–573.

[43] K. Unterrainer, R. Colombelli, C. Gmachl, F. Capasso, H.Y. Hwang, A.M. Sergent, et al., Quantum cascade lasers with double metal-semiconductor waveguide resonators, Appl. Phys. Lett. 80 (2002) 3060–3062.

[44] G. Sun, R.A. Soref, H.H. Cheng, Design of a Si-based lattice-matched room-temperature GeSn/GeSiSn multi-quantum-well mid-infrared laser diode, Opt. Express 18 (2010) 19957–19965.

[45] M. Liu, X. Yin, E. Ulin-Avila, B. Geng, T. Zentgraf, L. Ju, et al., A graphene-based broadband optical modulator, Nature 474 (2011) 64.

[46] S.J. Koester, M. Li, High-speed waveguide-coupled graphene-on-graphene optical modulators, Appl. Phys. Lett. 100 (2012) 171107.

[47] A. Vakil, N. Engheta, Transformation optics using graphene, Science 332 (2011) 1291.

[48] A. Pospischil, M. Humer, M.M. Furchi, D. Bachmann, R. Guider, T. Fromherz, et al., CMOS-compatible graphene photodetector, Nat. Photonics 7 (2013) 892.

[49] X. Gan, R.-J. Shiue, Y. Gao, I. Meric, T.F. Heinz, K. Shepard, et al., Chip-integrated ultrafast graphene photodetector with high responsivity, Nat. Photonics 7 (2013) 883–887.

[50] R.R. Nair, P. Blake, A.N. Grigorenko, K.S. Novoselov, T.J. Booth, T. Stauber, et al., Fine structure constant defines visual transparency of graphene, Science 320(5881) (2008) 1308.

[51] K.F. Mak, J. Shan, T.F. Heinz, Seeing many-body effects in single- and few-layer graphene: observation of two-dimensional saddle-point excitons, Phys. Rev. Lett. 106 (2011) 046401.

[52] P. Avouris, S. Member, M. Freitag, Graphene photonics, plasmonics, and optoelectronics, IEEE Journal of Selected Topics in Quantum Electronics. 20 (2014) 6000112.

[53] H. Li, Y. Anugrah, S.J. Koester, M. Li, Optical absorption in graphene integrated on silicon waveguides, Appl. Phys. Lett. 101 (2012) 111110.

[54] B. Scharf, V. Perebeinos, J. Fabian, P. Avouris, Effects of optical and surface polar phonons on the optical conductivity of doped graphene, Phys. Rev. B87 (2013) 035414.

[55] E.J.H. Lee, K. Balasubramanian, R.T. Weitz, M. Burghard, K. Kern, Contact and edge effects in graphene devices, Nat. Nanotech. 3 (2008) 486–490.

[56] F. Xia, et al., Photocurrent imaging and efficient photon detection in a graphene transistor, Nano Lett. 9 (2009) 1039–1044.

[57] L. Ju, B. Geng, J. Horng, C. Girit, M. Martin, Z. Hao, et al., Graphene plasmonics for tunable terahertz metamaterials, Nat. Nanotechnol. 6 (2011) 630–634.

CHAPTER 4

Moore's Law for Photonics and Electronics

DOWNSCALING OF CMOS

During the last five decades, semiconductor industry has experienced a continuous downscaling process of the silicon CMOS technology and today a struggle to achieve beyond 14 nm technology node is ongoing. This strategy is followed according to a roadmap which was proposed

Monolithic Nanoscale Photonics—Electronics Integration in Silicon and Other Group IV Elements.
DOI: http://dx.doi.org/10.1016/B978-0-12-419975-0.00004-0
© 2015 Elsevier Ltd. All rights reserved.

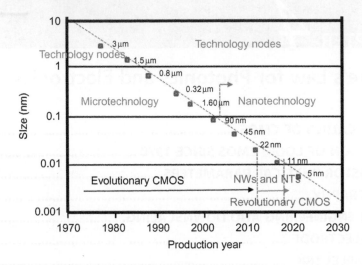

Figure 4.1 Transistor gate length in technology nodes and production year.

in 1965 by Intel's founder, G.E. Moore [1]. The roadmap requires a technology development to double the number of transistors per chip every 18 months. The main effort behind this downscaling process is to increase package density and circuit speed but also to lower dynamic power dissipation and cost-effective manufacturing. For CMOS, this means that the entire technology for device manufacturing (including lithography, etch, and epitaxy) has to be continuously updated as well.

The technology node roadmap results in a change of electrical and physical parameters of transistors. This urges both the lateral and vertical parts of the transistors, including the dopant profiles have to be modified.

In order to track the technological development in the United States, National Technology Roadmap for Semiconductors (NTRS) was organized to prepare a guideline for American semiconductor industry. This organization was reformed and International Technology Roadmap for Semiconductors (ITRS) was built in 1998, which consists of experts from the United States, Europe, South Korea, Japan, and Taiwan. Figure 4.1 shows the technology roadmap for Intel's processors for the introduction year to the market.

In order to further reduce the manufacturing costs, the downscaling of transistors was followed by increasing the size of Si wafers. The wafer size has been increased from 75 mm in 1970 to 300 mm in 2008 and 450 mm is available now in 2014 as shown in Figure 4.2.

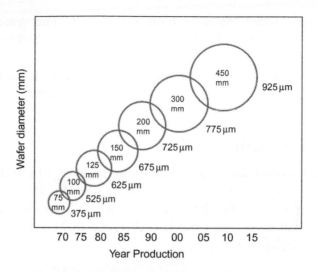

Figure 4.2 Wafer size and thickness in fabrication since 1970 [2].

The choice of batch-wafer or single-wafer processing was a topic for semiconductor industry for years. The derived conclusions suggested that a single-wafer solution is easier and more reliable in terms of uniformity and layer quality in process steps.

The motivation for the shift to the larger wafer sizes is to reduce the cost of fabrication per die. For example, converting from 200 to 300 mm wafers saves one-third of the cost for the fabrication of a die.

A simple approximation for the number of die per wafer (DPW) for a given wafer size is [3]

$$\text{DPW} = \frac{\pi d^2}{4A_{\text{die}}} - \frac{\pi d}{\sqrt{2A_{\text{die}}}} \qquad (4.1)$$

where A_{die} is the area of a die and d is the diameter of the Si wafer. The second term in Eq. (4.1) is a correction factor (circumference/die diagonal length) to area ratio in order to obtain a better estimation for the dies at the circumference of the wafers. There are other correction factors where the area ratio is scaled by an exponential and polynomial factor to obtain a better estimation for the DPW [3].

The mechanical and thermal handling are two critical issues concerning Si wafers with large sizes. The mechanical handling could be managed from 75 to 300 mm, but moving to 450 mm demands totally new machinery which needs a huge investment. The reason comes from the large volume of 450 mm wafers (925 μm thickness) which makes these wafers about three times heavier than 300 mm wafers. As a result, longer cooling time for different process steps is needed, which makes the whole process time considerably longer.

EVOLUTION OF LOGIC CMOS SINCE 1970

Downscaling started with 8 μm technology node in 70 s and extended to nm-scale transistors in 2003. An ultimate limit for this downscaling can be 0.3 nm which is the distance between two Si atoms, but the end of this path will be reached much earlier due to high cost and other practical limitations.

The downscaling refers to transistor's channel length (nm), density (gates/mm^2), dynamic capacity (gates), power (nA/gate), and cost (price/gate). Three technology node periods are addressed in the following sections [4]

Prior to NTRS and ITRS Roadmaps
Downscaling 8 μm > 6 μm > 4 μm > 3 μm > 2 μm > 1.2 μm > 0.8 μm > 0.5 μm

- The lithography resolution was the same as the gate length and half pitch of lines
- Downscaling of 0.7 μm per every 3 years
- Wafer size: 75–100 mm (1970–1980)

After NRTS and ITRS
350 nm > 250 nm > 180 nm > 130 nm > 90 nm > 65 nm > 45 nm > 32 nm > 22 nm (3D trigate)

- The gate length shrunk with one or two generations from the half pitch
- High-k was integrated in 45 nm node and beyond
- Integration of TiSi$_2$ in 180 nm, CoSi$_2$ in 65 nm, and NiSi in 45 nm nodes and beyond
- Wafer size: 200 nm (1990–2000) > 300 mm (2000–2010)
- Transition from 2D to 3D transistors (2011)
- More Moore and more than Moore

Future of Logic CMOS and Beyond CMOS
14 nm $>$ 10 nm $>$ 8 nm $>$ 5.5 nm

- Wafer size: 450 mm (? year)
- Fabrication of 3D transistors: (2011−?)
- Transition from Si to new channel materials
- Beyond CMOS

TRANSISTOR PHYSICAL PARAMETERS

The move to smaller sizes results in transistors with shorter channel length, functioning at lower supply voltages and switching with less current. The effects of shorter channel for transistors are lower gate capacitance and higher drive, where both leads to faster devices. At the same time, the shorter channels suffer from higher source−drain leakage. Gate leakage, as well, becomes larger due to the thinner gate oxide.

The other effects of downscaling are lower supply voltage (V_{DD}) and threshold voltage (V_T) which lead to lower dynamic power.

The main principles of conventional downscaling of MOSFET structure with a factor of γ is shown in Figure 4.3 [5].

When the physical dimensions of a transistor (gate length and width, oxide thickness, junction depth, and substrate doping) are downscaled, then the actual voltages (supply and threshold voltages) are also scaled by factor of γ. In this way, the electric field is kept constant and it is as in the original transistor, but at the same time the

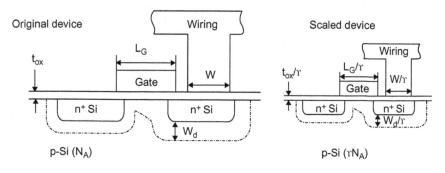

Figure 4.3 A schematic sketch of conventional downscaling of a MOSFET.

Table 4.1 Scaling Rules' Factors for Physical Parameters of MOSFETs

Physical Parameters	Scaling Rule's Factors		
	Constant Field	Generalized	Selective
Gate length (L_G), oxide thickness (t_{ox})	$1/\gamma$	$1/\gamma$	$1/\gamma_D$
Wiring width, channel width (W_G)	$1/\gamma$	$1/\gamma$	$1/\gamma_W$
Voltages (V_{DD}, V_T)	$1/\gamma$	β/γ	β/γ_D
Substrate doping (N_A)	γ	$\beta\gamma$	β/γ_W
Electric field	$1/\gamma$	β	β
Gate capacitance ($C = L_G W_G \varepsilon_{ox}/t_{ox}$)	$1/\gamma$	$1/\gamma$	$1/\gamma_W$
Drive current ($I_{D,sat}$)	$1/\gamma$	β/γ	β/γ_W
Intrinsic delay ($\tau \sim C_{VDD}/I_{D,sat}$)	$1/\gamma$	$1/\gamma$	$1/\gamma_D$
Area ($A \propto L_G W_G$, or $\propto W_G^2$)	$1/\gamma^2$	$1/\gamma^2$	$1/\gamma_W^2$
Power dissipation ($P \sim {}_{ID,sat} V_{DD}$)	$1/\gamma^2$	β^2/γ^2	$\beta^2/(\gamma_W\gamma_D)$
Power density (P/A)	1	β^2	$\beta^2/(\gamma_W/\gamma_D)$

density of transistors in the chip is increased by γ^2. In this concept, the ratio between gate length and width is kept constant.

In a more practical way, γ can be considered as scaling factor for the transistor's physical dimensions, but the supply voltage is being downscaled by β/γ. This approach is known as generalized scaling [6]. If the downscaling is performed without the condition of constant ratio between gate length and width, then it is known as selective scaling. In both generalized and selective scaling, the dissipated power is downscaled instead by β^2. The presentation of these scaling methods for field effect transistors is given in Table 4.1 [5].

Figure 4.4 shows how downscaling of the channel length can affect transistor's power supply and threshold voltage as well as the gate oxide thickness. When the gate length is downscaled below $0.1\ \mu m$, several leakage mechanisms or SCE become significant and increase the static power dissipation per device and unit area [8]. SCE can be held under control by downscaling the vertical dimensions, e.g., gate oxide thickness and junction depth.

The CMOS transistors function as a digital switch in CMOS logic circuits where the node capacitance is charged by a PMOS transistor and then discharged by NMOS one.

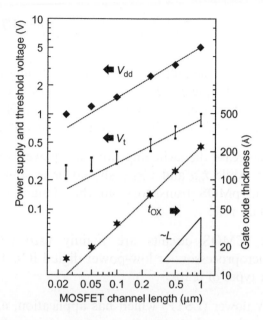

Figure 4.4 The dependence of power supply voltage, threshold voltage, and gate oxide thickness (t_{ox}) to CMOS channel length [7].

In this case, it is very important that the time of charge/discharge should be small for the capacitor. This may occur when the current for charging and discharging is high and the loading capacitance is small. In order to obtain a fast switching performance, all the parasitic capacitances of the junctions and gate have to be minimized. In such a situation, the drive current of transistors during the switching action (both in the saturation and linear regime) has to be high, while the series resistances and carrier mobility should be low and high, respectively. This type of logic design can be considered as a ring oscillator where the delay time is given by [9]

$$\tau \propto C_{\text{OUT}} \frac{V_{\text{DD}}}{I_{\text{On}}} \qquad (4.2)$$

where C_{OUT} is the output capacitance, V_{DD} stands for the power supply, and I_{On} denotes the transistor drive current. Then the active and standby switching powers are obtained from relations of current, where $V_{\text{DD}} = V_{\text{GS}}$:

$$P_{\text{ac}} = C_{\text{OUT}} V_{\text{DD}}^2 f \qquad (4.3)$$

$$P_{off} = W_{TOT} V_{DD} I_{off} = W_{TOT} V_{DD} I_0 \exp\left(\frac{-qVT}{mKT}\right) \qquad (4.4)$$

where P_{ac} and P_{off} are the active and standby switching power, respectively, f is the clock frequency, I_{off} is off-current in the device, I_0 is the extrapolated current at a certain threshold voltage, m is ideality factor which is in range of ~ 1.2, and V is a given threshold voltage value [10].

In order to decrease the switching time and power, the I_{on} currents have to be high, while I_{off} and C_{OUT} should be as low as possible in both nMOS and pMOS transistors. In this design, therefore, V_{DD} plays an important role.

In principle, CMOS circuits are mainly used either in high-performance microprocessors or low-power digital ICs. The latter ones have two major types of chips:

1. Low standby power (LSTP) which has application, e.g., in cellular phones where lower battery consumption is intended. The chips contain severely downscaled transistors for high-performance ICs and high leakage current.
2. Low operating power (LOP) with relatively low-performance applications, e.g., portable computers or laptops. The transistors in the chips have lower performance and remarkably lower leakage current.

This means that downscaling of MOSFETs is important matter for LSTP and telecom application, where the low-power battery is an essential issue.

LITHOGRAPHY

Gate lithography is an important issue in CMOS downscaling. The main way in achieving smaller feature sizes is to use light sources of shorter wavelengths in lithography. Further improvement in resolution can be obtained by different resolution enhancement techniques as shown in Figure 4.5.

The first generation of light source in lithography was provided by Hg lamps to generate 400 nm wavelength with spectral lines at 436 nm ("g-line"), 405 nm ("h-line"), and 365 nm ("i-line").

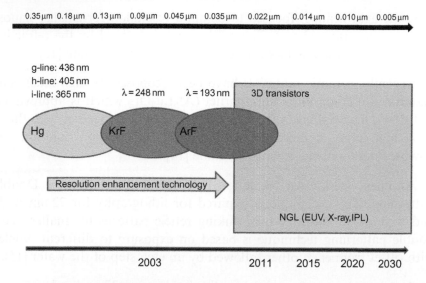

Figure 4.5 Lithography techniques in downscaling process.

Downscaling roadmap demanded higher resolution together with higher throughput to achieve the goals to increase the number of transistors in chip and lower the production cost. Eventually, the lithography tools with Hg lamp as light source could not fulfill these industry's requirements.

The next generation of lithography light source became deep ultraviolet (DUV) excimer lasers. These types of lasers range from the krypton fluoride laser to generate 248 nm wavelength and the argon fluoride laser for 193 nm wavelength [11].

In order to enhance the lithography technique optical proximity correction (OPC) is commonly applied to compensate for image inaccuracies caused by diffraction [12]. The projected features appear with irregularities in line widths and rounded corners. These distortions may affect the electrical properties of the manufactured devices. The compensation is done by changing the pattern on the mask through moving edges or introducing extra polygons to the pattern in the mask. As help tools, precomputed tables based on size and density of features or simulating the final pattern to find out the movement of edges are utilized to obtain the best solution.

Immersion lithography is a technique where air between the lens and the wafer is replaced with purified liquid with refractive index >1. In this way, the light beam is focused to obtain a resolution for smaller nanos-caled features (45−32 nm). The resolution is improved by the refractive index of water which is 1.44 for light with 193 nm wavelength [13].

Recently, ceramic materials have been proposed as high-index lens material. Lutetium aluminum garnet ($Al_5Lu_3O_{12}$ which is abbreviated to LuAG) material has refractive index of 2.14 and together with fluids with high-index immersion constitute an outstanding solution for downscaling to 10 nm technology node [14].

Another well-known resolution enhancement technique is Double patterning. This technique is required for lithography for 22 nm node and beyond. In contrary to shrinking reticle patterns to smaller sizes, double patterning technique is based on exposure to different reticles with offset from each other followed by an etch step of the wafer [15].

Another alternative to lithography for beyond 22 nm node is using light sources with much shorter wavelength, e.g., extreme ultraviolet (EUV) with 13.5 wavelength, X-ray lithography (XRL) with 1 nm wavelength [16] and deep X-ray lithography (DXRL) with 0.1 nm wavelength [17]. These techniques belong to next-generation lithography (NGL) encompassing the lithography technologies which replace the current photolithography.

EUV technique is still under development and is expected to be used up to 5.5 nm technology node. The main barrier to use this technique in fabrication is the tremendous high price.

The XRL has the complexity that the light source is noncollimated and requires special collimating mirrors and diffractive lenses instead of traditional refractive lenses. The masks for XRL are made of Ta, W compounds or gold as X-ray absorber, and are placed on a membrane of SiC which is transparent to X-rays.

Among lithography techniques, electron-beam lithography (EBL) has attracted attention in many research laboratories. EBL is a technique which is based on scanning of a focused electron beam to draw features on an electron sensitive resist. Since the size of the electron beam is small, this technique is capable to create usually lines of 20 nm resolution.

The advantage of EBL is being maskless with high resolution. However, this technique has a low output since the time for writing is relatively high. To expose electrons with a beam current of I to a certain area, A, a right energy dose, D, and exposure time, T, are required according to [18]: $T \cdot I = D \cdot A$.

One limitation with EBL technique is when small features are patterned the number of electrons in the beam has to be decreased. This leads to a large variation of dose due to a shot noise effect.

Although EBL offers a high resolution but some deformations may occur due to different backgrounds: data-related or physical-related defects. The latter problem is due to sample charging, backscattering, beam deflection, and beam drift.

Data-related defects is the condition that e-beam is not deflected accurately causing deformation in the shape or even blanking.

The new generation of EBL is based on the secondary electrons as primary beam. Secondary electron beam is induced when high-energy electrons kick out an electron from the outer shell of an atom. This type of electrons has remarkably lower energy and generates less beam-related defects.

STRAIN ENGINEERING AND DOWNSCALING

To strain the channel region has been the most essential technique to improve the carrier transport and boost the mobility. The original idea to induce strain in the channel was based on biaxial strain, mainly in Si. The presence of strain becomes more important when high-k materials are integrated as gate insulator and the interfacial imperfectness degrades the channel mobility. $Si_{1-x}Ge_x$ and $Si_{1-y}C_y$ have been used as stressor materials in the S/D regions [19]. These regions are etched first to form recess shape and the stressor materials are deposited to strain the Si in the channel. Nitride layers are mostly deposited on the transistor body to induce strain as well [19].

More details about the strain engineering in MOSFET structure is provided in Chapter 1, part two. Figure 4.6 shows the SiGe layers with different compositions for each technology node. The induced strain is proportional to the Ge content in SiGe layers. The strain amount is within a few Giga pascal (GPa) in the channel region of these transistors.

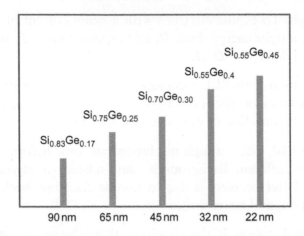

Figure 4.6 SiGe as stressor material in S/D regions from 90 nm (in 2003) to 22 nm (in 2011).

GATE ELECTRODE

Polycrystalline Si (or poly-Si) was a common gate electrode material in many MOSFET generations until 130 nm technology node. Integration of a poly-Si electrode in MOSFET is not free of difficulties, and a series of problems rises because of nonuniform doping profile and dopant penetration into the oxide. The former problem is due to insufficient doping in polycrystalline material close to the oxide interface. This problem results in a depletion region in the poly-Si side which appears in the form of an increase in the effective gate oxide thickness. The dopant penetration occurs for example in B-doped poly-Si when boron diffuses through the thin oxide layer. A solution to boron penetration problem is to introduce nitrogen into the oxide layer. Unfortunately, such nitrogenated gate oxides degrade the carrier mobility in the channel. Replacing poly-Si with a metal gate solves all these problems and search for different suitable candidates has been ongoing for years.

Selecting an appropriate metal gate material is very important for MOSFETs since V_T is affected by the metal workfunction [20]. Metal gates with workfunctions within 0.2 eV of the conduction and valence band edges of Si are required to replace n- and p-type poly-Si gates as shown in Figure 4.7. During the last few years, search to find suitable metal gate for CMOS technology has been intense and yielded several candidate metals with high melting point [21].

The research for more complicated metal stacks as gate electrode has been and is still ongoing. W/TiN, Mo, Ta, TaN, TiN Ti$_x$Al$_y$,

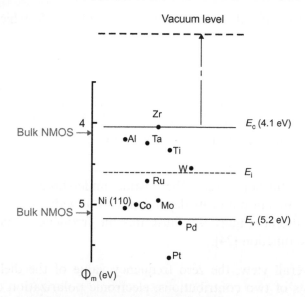

Figure 4.7 Gate electrode candidate materials for bulk NMOS and PMOS.

$Ti_xAl_yO_x$, and $TaSi_xN_y$ are the most outstanding metal gates demonstrated till now [22].

GATE DIELECTRIC

The most serious material change in the classical MOSFET structure addresses gate dielectric. The thickness of SiO_2 as gate dielectric has been scaled down from 300 nm in 10 μm node to 1.2 nm in 65 nm technology node. The continuous reduction of gate oxide below 1.2 nm results in increasing tunneling current and gate leakage current. It is worth noting here that the high-performance circuits may operate with higher gate leakage than the LOP circuits.

For example, when SiO_2 thickness is 1.5 nm, the leakage current density is estimated to be 100 A/cm^2 for 1.5 V source voltage [23]. However, any allowable thickness of SiO_2 is still too high for the demanded performance.

The solution to this essential problem is to find materials with higher value of dielectric constant (the so-called high-k materials) than SiO_2. In this way, the high-k material can be integrated as gate dielectric in CMOS structure with a thickness which is thicker than SiO_2 with a factor of εhigh-k/ε_{ox} for the desired gate capacitance.

For example, the dielectric constant of HfO_2 is 35, which is about 10 times higher compared to 3.9 for SiO_2. So, this dielectric material can be grown ~ 10 times thicker than SiO_2 to decrease the leakage current.

In general, the thickness of gate dielectric is expressed in equivalent oxide thickness (EOT). EOT for a high-k is defined as the thickness of silicon oxide layer which is needed to obtain the same capacitance density as for the high-k material. This value is obtained from EOT = $(3.9/k)t_{ox}$.

In order to further increase the k-value, understanding the nature of the dielectric function and its dependency on a wide range of frequencies is very useful. Figure 4.8 illustrates the frequency dependency of the dielectric function [24].

In an overall view, the zero frequency value of the dielectric constant consists of two contributions: electronic polarization component which is dominant at high frequency and the ionic contribution [5]. For CMOS applications, the frequency window lies in a region where both electronic and ionic contributions (ε_∞ and ε_{latt}, respectively) impact on the k-value. Therefore, the permittivity is given by the relation [25]

$$\varepsilon_{ox} = \varepsilon_\infty + \varepsilon_{latt} \tag{4.5}$$

In the above relationship, the electronic component relates to refractive index by $n \sim \sqrt{\varepsilon_\infty}$.

A way to increase k-value is applying materials with large lattice component. This means that depending on the crystalline forms ε_{latt} can be tailored and affect the k-value. For example, for different

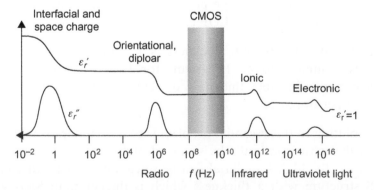

Figure 4.8 Frequency dependency of dielectric function.

crystalline phases of HfO_2, high-k values may vary from 1.82 to 26.17 [25]. It is emphasized here that for CMOS applications amorphous materials with high crystalline temperature are desired.

The k-value can be related to the polarizability (α) and the volume of the unit cell (V_m) through Clausius–Mossotti (C–M) theory as obtained from the following equation [26]:

$$\varepsilon_r = \frac{1 + \frac{2}{3} 4\pi \frac{\alpha}{V_m}}{1 - \frac{1}{3} 4\pi \frac{\alpha}{V_m}} \tag{4.6}$$

In principle, larger atoms result in more polarization and therefore higher k values can be obtained. The C–M equation shows that k-value rises sharply when the ratio α/V_m increases demonstrating a strong dependence on both structure and the nature of high-k material.

A range of research to grow high-k materials has been carried out for years. Figure 4.9 shows different dielectrics and their measured k-values.

Requirements of ITRS for 22 nm technology node are EOT = 0.5 nm and leakage current density $< 10^{-2}$ A/cm^2 at 1 V, considering different tunneling processes. To fulfill these requirements, only oxides HfO, Pr_2O_3 in the hexagonal phase, La_2O_3, and $LiNbO_3$ are found excellent but may be Sm_2O_3, Ce_2O_2, and Gd_2O_3 can be nominated as well.

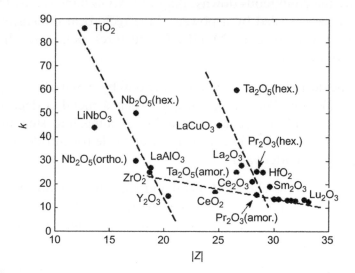

Figure 4.9 The variability of relative permittivity values of dielectrics in terms of their constituents' atomic number [26,27].

It is worth mentioning here that among the lanthanides, La_2O_3 is perhaps the most appropriate high-k material for future [28]. The crystalline temperature of HfO_2 can be enhanced by alloying with Al_2O_3 to form an HfAlO material. The thermal stability of amorphous HfAlO may be improved to at least 900°C [29,30]. Alternative to above aluminates are silicate alloys, where $(HfO_2)_x(SiO_2)_{1-x}$ are formed [24]. A novel material such as thulium silicate (TmxSiyOz) provides a robust template to grow any other k-high stack on Si. TmSiO material is formed through annealing of the deposited layer at 550−600 C. The thickness of the layer is dependent of the annealing temperature and the phase is formed within initial seconds (self-limiting process). The measured interface state density was within $\sim 0.7 - 2 \times 10^{11}$ cm^2eV^{-1} at flat-band condition, where subthreshold slopes of ~ 70 mV/dec were obtained for pFET and nFET [31].

So far, hafnium oxide (HfO_2), hafnium silicate (HfSiON), and zirconium oxide (ZrO_2) are most mostly used as gate dielectrics [32].

These materials with $\kappa \sim 10-15$ compared to 3.9 for SiO_2 may decrease the gate leakage currents significantly.

CONTACT RESISTANCE

The silicides are formed through a chemical interaction of a metal with silicon. By the continuous downscaling of CMOS dimensions, the contact resistance is scaled by a power of the reciprocal dimensions [33]. The contact resistance in MOSFET structures points to the gate, source, and drain contacts.

The silicide is formed over these regions by a self-aligned process called salicidation. This process occurs when a metal is deposited on the transistor and silicide is formed by an annealing step in direct contact with silicon. The excess metal over oxide (or nitride) surface is selectively etched afterward.

Depending on the choice of metal and annealing conditions, a high- or low-resistivity phase of silicide can be formed after the first RTA. A second RTA is then required to transform the high-resistivity phase into the required low-resistivity phase [34].

An appropriate silicide is desired to be formed at low temperature range and consume low amount of Si. The first generation of silicide layers was formed by using Ti, Co, and Pt metals.

Silicide	Formation Temperature (°C)	Resistivity (μΩ cm)	Silicon Consumed (Å) per Å of Metal	Resulting Silicide Thickness (Å) per Å of Metal	Φ_b on n-type (eV)
$TiSi_2$	800–900	13–16	2.27	2.51	0.6
$CoSi_2$	600–700	18–20	3.64	3.52	0.64
PtSi	300–600	28–35	1.32	1.97	0.87
NiSi	400–600	14–20	1.83	2.34	0.7

Table 4.2 Most Common Silicide Materials with Their Material Properties

In 65 nm technology node and beyond, Ni was chosen as an outstanding metal for silicides for CMOS structure [35]. NiSi has several advantages compared to other silicides, e.g., $TiSi_2$ and $CoSi_2$ including low sheet resistance on narrow lines as well as low-temperature process with low silicon consumption as given in Table 4.2 [36]. The latter is important for the formation of contacts to ultra shallow S/D junctions.

SUBSTRATE DESIGN

The substrate can be designed according to the demanded application. Bulk (450 mm is available today) and SOI wafers (300 mm ultrathin SOI is available today) in different sizes can be used for electrical and photonic devices. Ultrathin body (UTB) SOI substrates have been used as substrate to manufacture fully or partially depleted MOSFETS. The main goal is to improve the speed since all parasitic capacitances can be removed [37].

The Si in UTB SOI is usually lightly doped and this facilitates a low charge density in depletion operation mode. As a result, it is difficult to change V_T. The doping level can be increased, but the carrier mobility will be degraded. In UTB SOI transistors, the thickness of Si may affect V_T as well. However, the uniformity of Si becomes a real problem for ultrathin layers (<10 nm). Therefore, metal gate engineering becomes a vital subject for transistors manufactured on SOI [38]. In this case, in order to fulfill the specifications from ITRS for off-current, a metal gate with high workfunction and lowly doped Si channel is an excellent design for MOSFETs.

The recent developments in substrate engineering have provided new monolithic solutions for electronics and photonics.

For example, 3D integration of waveguides can provide the possibilities for advanced Si-based integrated chips. Dual waveguiding in Si at 1.33–1.55 μm in double SOI substrate (SiO_2–Si/SiO_2–Si/SiO_2

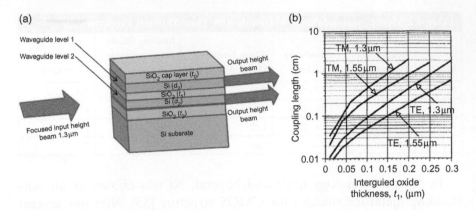

Figure 4.10 (a) A multilayer of Si–SiO₂ structure for vertical 3D integrated waveguides and (b) calculated coupling length versus t_1, for dual waveguiding in Figure 4.10(a) with $t_o = 1$ μm, $d_1 = d_2 = 2$ μm, and $t_2 = 0.37$ pm.

structure) is a possible example (Figure 4.10(a)). The thickness of the layers has an important role in how the dual waveguides function. If the thickness of intercore oxide is below 5—50 nm the guiding is leaking, meanwhile above 50 nm a normal waveguiding in the Si cores occurs with a strong mutual coupling. A thick intercore oxide results in waveguiding of light independently. Figure 4.10(b) demonstrates calculated curves for transition metal (TM) and TE polarization at the wavelengths of 1.3—1.55 μm in terms of t_1 [39].

There are also other 3D substrates such as a double SiGeOI [40] or even double GOI 3D substrates which can be used for photonic applications depending on the wavelength range. [41]. The lower bus is a place for both electronic and photonic components, and upper can be used for waveguides which transport the laser light to other components.

HEAT PRODUCTION

As the downscaling of MOSFETs is set forward to increase the density of MOSFETs in the chips, the heat generation becomes a severe issue. The circuits function more slowly at high temperatures.

The on-state resistance increases with temperature and when a constant-current load is applied to the transistors, the power loss grows accordingly, leading to more heat generation. The generated heat may damage the transistors causing reliability and lifetime issues. Nowadays, cooling devices and heat sinks are integrated for various ICs.

SHORT CHANNEL EFFECTS

One of the fundamental problems with downscaling of MOSFETs is the SCE. SCE occurs when the channel length is close to the depletion width of source and drain. It comprises the following effects: decrease of threshold voltage, increase of the slope of subthreshold voltage when the gate length is decreased, and the threshold voltage becomes dependent on the drain voltage (this is known as DIBL). All these will affect the off-state current and the transistor's switching property. In this situation, the transistor cannot be switched off well.

SCE depends on the closeness of source and drain, doping concentration in the channel region, and the extension of junction under the gate. In this case, the steepness of S/D extension has an important role in SCEs for nanoscaled transistors. The best way to counteract the SCE is optimizing the shallow junctions and a channel region with a super steep retrograde doping profile in addition to halo implantation technique [42].

A super steep retrograde channel profile is attributed to a channel region which is lowly doped close to the oxide gate, but within a few nm below the gate, dopant concentration rises steeply. This type of doping profile is excellent to provide a high surface mobility.

The halo doping at the source side may intersect with the drain side for the transistors with short channel length leading to an increase of the total channel doping level. This may result in an increase of the threshold voltage. As a consequence, a larger gate voltage is needed to obtain channel inversion.

While the control of SCE is a serious issue for 2D transistors, it becomes entirely different for 3D transistors. For example, multi-gate field effect transistor (MuGFET) can be scaled to 10 nm without any appearance of SCE. MuGFET devices are FinFET, trigate (two-gate transistors), and the gate-all-round transistors. The MuGFETs have considerably higher drive current compared to traditional MOSFETs. It has been shown that the SCE can be suppressed when the physical size (width and height) of trigate MOSFET is less than the effective gate length.

The SCEs in MOSFETs are classified to DIBL, punch through, mobility degradation, velocity saturation, and hot electron.

DRAIN-INDUCED BARRIER LOWERING

This is an effect in MOSFETs when the drain voltage can vary the output conductance and threshold voltage of the transistor. DIBL occurs for transistors where the gate length is downscaled without correctly scaling the other physical dimensions. In short channel MOSFETs, the depletion regions of source-body and drain-body p–n junctions are extended under the gate and compete with the gate role over the carrier transport as the drain voltage is increased. Then, the charge present under the gate is maintained in balance by drawing more carriers into the channel, a phenomenon which corresponds to the lowering of the threshold voltage of the MOSFET.

DIBL increases when the channel length is downscaled in different generations of technology nodes and it appears in the form of threshold modification for MOSFETS. For transistors with enormously short channel length, the gate may not be able to turn off the transistor.

DIBL is effectively reduced in FinFETs due to higher geometric control of the gate over the channel as the depletion regions are constrained by the fin itself and do not spread into the bulk. DIBL effect is one of the major issues which varies the threshold voltage in FinFETs. Figure 4.11 shows different fin profiles adapted for a constant DIBL [43].

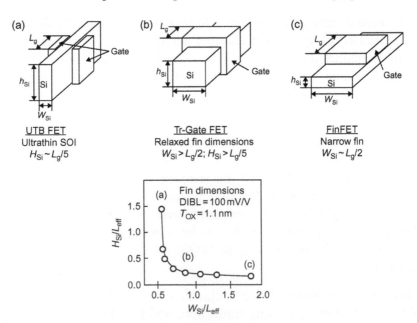

Figure 4.11 Dimension ratios for a constant DIBL effect in a FinFET.

DIBL can usually be removed by accurately scaling the drain and source junction depths and increasing doping concentration in the substrate of MOSFETs.

PUNCH THROUGH

Punch through is addressed to MOSFETs' channel length modulation and occurs when the depletion regions of the drain-body and source-body junctions meet and form a single depletion region. In this situation, the current flow in the channel is not controlled by the gate voltage and promptly increases with increasing drain-source voltage (Figure 4.12). This main reason for punch through in the transistors is the current transport occurs deeper in the bulk and far away from the gate. Then the subthreshold leakage current is increased resulting in an increased power consumption. The amount of punchthrough current is dependent largely on the applied drain voltage and on the source/drain junction depths.

A solution to decrease the punchthrough effect is to increase the doping level in transistor body (bulk). In this case, the drain and source depletion regions will be reduced and no parasitic current path is formed. However, a higher doping in transistor body increases the subthreshold swing, therefore, this method is not recommended to be used to reduce drain-source leakage.

Other methods to avoid punch through are to use of spatially restricted dopant implantations e.g. (a) halo, or pocket implantations and (b) delta doping (Figure 4.13).

MOBILITY DEGRADATION

The carrier transport in the channel region is crucial for the transistor characteristics. The carrier mobility in the inversion layer is lower than

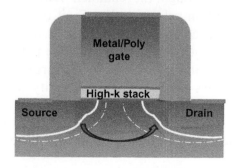

Figure 4.12 Schematic view of a MOSFET. Large drain bias can result in "punch through."

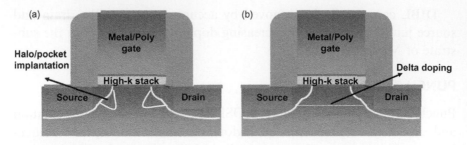

Figure 4.13 Methods to decrease punch through effect in MOSFETs.

in bulk material. The reason for this behavior is that usually a part of electron wavefunction penetrates into the oxide and the carrier mobility is degraded because of the lower mobility in the oxide. In nanoscale MOSFETs, the high electric field at the surface drives the electron wavefunction even further into the oxide region resulting in a field-dependent mobility. The surface mobility (μ_{sur}) varies with the electric field (ε) according to the following relation: $\mu_{sur} \propto \varepsilon^{-1/3}$.

VELOCITY SATURATION

Velocity saturation refers to the situation when the carrier velocity achieves a maximum value when a high electric field is applied (typically in the range of 10–100 kV/cm). At this point, the carrier velocity does not increase linearly as the applied electric field increases and the carriers release their access energy through interacting with lattice atoms and by emitting optical phonons. Velocity saturation plays an important role in voltage transfer characteristics of the transistors since at saturation velocity they do not follow Ohm's law. For MOSFETS, this effect results in an increase of the carrier transit time through the channel. Different semiconductors have different saturation velocities. For Si, it is $\sim 1 \times 10^7$ cm/s, but it can vary in the presence of crystal defects. In nanoscale MOSFETs, the size of channel length is close to mean free path of electrons. The average electron velocity in the channel is larger than bulk, so carrier transport is less restricted to the velocity saturation. There is a hope in integration of new channel materials, e.g., graphene or graphene-like materials which can demonstrate a ballistic transport. In such a case, the distance of S/D contacts will be smaller than the mean free path.

HOT ELECTRON EFFECT

Hot carrier injection in MOSFETs occurs when a carrier from Si channel is injected into the gate oxide. For this transition, a carrier should have a high kinetic energy to reach the conduction or valence band in the oxide. This energy amount for an electron and hole is 3.2 and 4.6 eV, respectively.

Hot electrons are usually generated when photons with high energy are shined on the transistors. The gained energy from the photon is transferred to an electron which is ejected out of the valence band and generates either an electron–hole pair or surpasses even the conduction band to become a hot electron.

Hot electrons are also generated when the channel is conductive and some electrons may get injected into the gate region and become hot. These electrons create a leakage current instead of flowing the current through the channel region. This leakage current may break an atomic bond, e.g., Si–H bond and even damage the dielectric material. As a consequence, the interface states may get increased which in turn degrade the channel mobility, modify the threshold voltage, degrade the subthreshold slope, and affect switching characteristics of the MOSFET.

Attempts to correct or compensate for the hot electron effect in a MOSFET may involve locating a diode in reverse bias at gate terminal or other manipulations of the device structure (such as lightly doped or double-doped drains).

3D CHIPS, NEW VISION FOR DOWNSCALING

The purpose to follow Moore's roadmap is downscaling the transistor dimensions and to increase the number of transistors. A serious attempt to end the miniaturization can be taken out if the number of transistors could be increased in three dimensions as shown in Figure 4.14. Each discrete chip is called a tier which is piled on each other and being packed. The inter-contacts can be formed among tiers by vertical copper pillars in vias. The idea of sequential fabrication or 3D circuitry is absolutely not new. To build additional packed layers of transistors on top of the first one may extend the Moore's concept to remarkably longer time. This design does not demand manufacturing smaller transistors in order to enhance the performance of ICs.

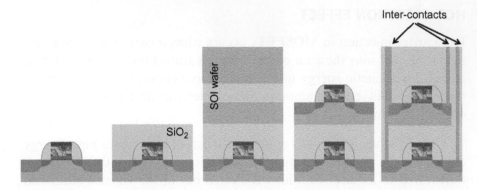

Figure 4.14 Processing steps of a 3D chip.

The other way to manufacture 3D chips is to bond two different processed wafers. The bonding process has to be performed at a temperature as low as possible to avoid any damage to the transistors.

The main disadvantage with 3D chips is the handling of the localized heat production. The performance of transistors inside such chips may get degraded since the structure profile and contacts are sensitive to high temperature. Advanced heat sink methods are required to control the heat production.

DOWNSCALING FOR NEXT 30 YEARS

Now, the downscaling has reached to an era which is more difficult than ever, experiencing a remarkable increase of cost in lithography, vast concerns for emerging new technologies, and difficulties in stability of electrical characteristics of MOSFETs in nm geometry. Nevertheless, it is expected that the downsizing will remain on the route by all means toward all technological boundaries for several generations during another 20–30 years, although the time between consecutive generations would turn out to be longer. Some of the visions for development of logic CMOS transistors and other devices have been shown in Figure 4.15.

As it was discussed before, the main limiting factors for downscaling are the parasitic resistances and capacitances. A real strong effort is needed to solve these problems in addition to reducing the costs of lithography. Moving to 450 mm wafer is a real and difficult decision for the chip manufacturers. How to handle these wafers in processing

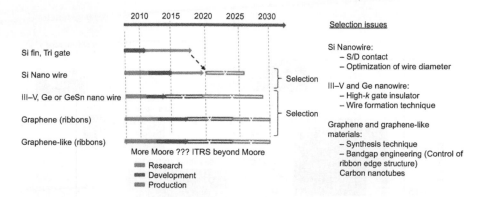

Figure 4.15 Roadmap for CMOS transistors in future [4].

is an obstacle to use these wafers in mass production. The choice of future channel material is still fuzzy too. The criteria for an excellent channel material are high mobility and density-of-states for electron and holes. The latter requirement is important since it has influence on the inversion charge as well as the drive current. However, many of the high-mobility channel materials suffer from low density-of-states. As an example, carbon nanotubes have several orders of magnitude lower DOS than Si, which makes it difficult to obtain high drive current out of them.

Nowadays, it is an attitude to move forward to Ge material but still passivation and high-k material integration are some critical issues. It seems that GeSnSi is an excellent buffer layer (or template) between Si and III−V materials. This paves the path for III−V HEMT application which has a huge importance for transistor technology. Graphene and graphene-like materials are still in research/ development, but it is believed that advanced transfer methods will provide a reliable way to create 2D single- or multi-sheets on 300 or 450 mm wafers. In this case, we may use graphene in both photonic and electronic applications solely or in combination with Si-based devices in future.

MOORE'S LAW FOR INTEGRATED PHOTONICS DEVICES AND SOME VISION FOR THE FUTURE

While the total number of devices in integrated photonics chips is still on the order of hundreds depending on how one counts, that of the

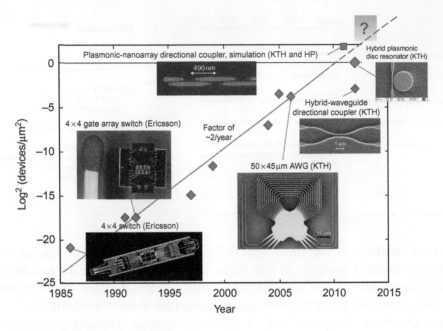

Figure 4.16 A Moore's law for integration density in terms of equivalent number of elements per square micron of integrated photonics devices, showing a growth faster than the IC Moore's law. The figure covers, in time order, a lithiumniobate 4×4 polarization-independent switch array, a 4×4 InP-based integrated gated amplifier switch array, an SOI arrayed waveguide grating, and a hybrid plasmonic (passive) directional coupler. All these are experimentally demonstrated. At the top is a simulation of two coupled metal nanoparticle arrays, forming a directional coupler, each array being a resonantly operated array of silver nanoparticles. If loss requirements of, for example, 3 dB/cm were invoked, the two latter would occupy significantly lower places in the figure. Adapted from [46].

electronic IC is on the order of billion transistors with an exponential growth in accordance with the famed Moore's law [1], a law which has turned out to be a formidable prediction of the future, or maybe formed the future, and a law which maybe has more economic than technological ramifications. In some contrast to the development of the number of devices on a photonics chip, the integration density has seen an exponential growth of the same order as that of electronics [44]. We have thus endeavored to formulate a photonics version of Moore's law, Figure 4.16, based on integration density, which we define in terms of equivalent functions, since we do not have generic components, such as transistors and resistors in photonics. As an example, we can take an 8 × 8 arrayed waveguide grating (AWG) [45], which in view of its functionality cannot reasonably be regarded as one single element. Difficulties of this nature do not appear in integrated electronics.

Some differences between the two types of ICs might be worthwhile to point out:

a. Moore's law for electronic ICs pertains to circuits with generic elements (transistors, resistors, capacitors), some fraction of which are operative in the sense that they dissipate power. These elements are fabricated by standard processes, applicable to all elements, basically in one material, silicon (though this is somewhat changing) with its natural passivating oxide.
b. A similar "Moore's law" for photonics will have to take into account the fact that no generic elements like in electronics exist, on the contrary, the elements are different, employ differing fabrication processes, and the materials are different (III−V semiconductors, silicon, ferroelectrics, polymers, etc).
c. There is no or small power dissipation in the passive fabric case, such as in the AWGs and switch arrays in ferroelectric, and there is "high" power dissipation for active devices (lasers, optical amplifier) and intermediate in high-speed modulators.

The exponential development in integration density has been made possible by several factors: materials with higher refractive indices, such as going from quartz ($n = 1.5$) to silicon ($n = 3.5$), with III−V compounds of slightly lower index as an intermediate step (see Figure 4.16). High index contrast in optical waveguides allows stronger light confinement and smaller bending radii and in general more efficient components [47]. Another issue is increased fabrication precision and improved modeling tools.

REFERENCES

[1] G.E. Moore, Electrical Engineer, an oral history conducted in 1976 by Michael Wolff, IEEE History Center, Rutgers University, New Brunswick, NJ.

[2] Hideki Tsuya, Present Status and Prospect of Si Wafers for Ultra Large Scale Integration, Jpn. J. Appl. Phys. 43(7A) (2004) 4055−4067.

[3] D.K. de Vries, Investigation of gross die per wafer formulas, in: IEEE Transactions on Semiconductor Manufacturing, 2005, pp. 136−139. Available from: http://dx.doi.org/10.1109/TSM.2004.836656.

[4] H. Iwai, Roadmap for 22 nm and beyond, Microelectronic Eng. 86 (2009) 1520.

[5] D.J. Frank, R.H. Dennard, E. Nowak, P.M. Solomon, Y. Taur, H.-S. Philip Wong, Device scaling limits of Si MOSFETs and their application dependencies, Proc. IEEE 89 (2001) 259.

[6] G. Baccarani, M. Wordeman, R. Dennard, Generalised scaling theory and its application to a ¼ micrometer MOSFETdesign, IEEE Trans. Electron. Dev. 31 (1984) 452.

[7] Y. Taur, CMOS scaling beyond 0.1 p: How Far Can It Go? International Symposium on VLSI Technology, Systems, and Applications, Proceedings of Technical Papers., Taiwan, Taipei, (1999) 6.

[8] S.-H. Lo, D.A. Buchanan, Y. Taur, W.I. Wang, Quantum-mechanical modeling of electron tunneling current from the inversion layer of ultra-thin-oxide nMOSFET's, IEEE Electron. Device Lett. 18 (1997) 209.

[9] Y. Taur, T.H. Ning, Fundamentals of modern VLSI devices, Cambridge University Press, Cambridge, United Kingdom (2013). ISBN-10: 0521559596.

[10] Y. Taur, CMOS design near the limit of scaling, IBM I. Res. Dev. 46 (2002).

[11] D. Williamson, DUV or EUV, that is the question, Proc. SPIE, (2000) 4146.

[12] A. Wong, Resolution Enhancement Techniques in Optical Lithography, SPIE Press, Bellingham, Washington, USA, 2001.

[13] H. Sewell, P. Graeupner, D. McCafferty, L. Markoya, N. Samarakone, P. van Wijnen, et al., An update on the progress in high-n immersion lithography, J. Photopolymer Sci. Technol. 21 (2008) 613.

[14] Y. Liberman, M. Rothschild, S.T. Palmacci, R. Bristol, J. Byers, N.J. Turro, et al., High-index immersion lithography: preventing lens photocontamination and identifying optical behavior of LuAG, in: Proc. of SPIE, 6924 (2008) 692416. Available from: http://dx.doi.org/10.1117/12.771462.

[15] M. Levinson, Double double, toil and trouble, Microlithography World, article 286361, 2007.

[16] Y. Vladimirsky, A. Bourdillon, O. Vladimirsky, W. Jiang, Q. Leonard, Demagnification in proximity x-ray lithography and extensibility to 25 nm by optimizing Fresnel diffraction, J. Phys. D: Appl. Phys. 32(22) (1999) 114.

[17] Deep X-ray lithography (DXRL) with 0.1 nm wavelength.

[18] N.W. Parker, A.D. Brodie, J.H. McCoy, High-throughput NGL electron-beam direct-write lithography system, in: Proc. SPIE 3997, 2000, 713. Available from: http://dx.doi.org/10.1117/12.390042.

[19] R. Arghavani, L. Xia, H. M'Saad, M. Balseanu, G. Karunasiri, A. Mascarenhas, S.E. Thompson, et al., A reliable and manufacturable method to induce a stress of > 1 GPa on a P-channel MOSFET in high volume manufacturing, IEEE Electron. Device Lett. 27 (2006) 114.

[20] Y.-C. Yeo, Metal gate technology for nanoscale transistors—material selection and process integration issues, Thin Solid Films 462−463 (2004) 34−41.

[21] Chang-Hoon, et al., Gate length dependent polysilicon depletion effects, IEEE Electron. Device Lett. 23 (2002) 224−226.

[22] A. Farkhanda, U.-D.Najeeb, Gate workfunction engineering for deep sub-micron MOSFET's: motivation, features and challenges, Int. J. Electron. Commun. Technol. 2 (2011) 29.

[23] Semiconductor Industry Association International Technology Roadmap for Semiconductors, 2001.

[24] G.D. Wilk, R.M. Wallace, J.M. Anthony, High-k gate dielectrics: current status and materials properties considerations, J. Appl. Phys. 89 (2001) 5243−5275.

[25] G.-M. Rignanese, Dielectric properties of crystalline and amorphous transition metal oxides and silicates as potential high-k candidates: the contribution of density-functional theory, J. Phys. Cond. Matt. 17 (2005) R357−R379.

[26] O. Engstrom, B. Raeissi, S. Hall, O. Buiu, M.C. Lemme, H.D.B. Gottlob, et al., Navigation aids in the search for future high k dielectrics: physical and electrical trends, Solid-State Electron. 51 (2007) 622–626.

[27] S. Hall, et al., Review and perspective of high-k dielectrics on silicon, J. Telecomm. Inf. Technol. 2 (2007) 33–43.

[28] H. Iwai, S. Ohmi, S. Akama, C. Ohshima, A. Kikuchi, I. Kashiwagi, et al., Advanced gate dielectric materials for sub-100 nm CMOS, in: Int. Electron Dev. Meet. Tech. Dig., 2002, pp. 625–628.

[29] M. Cho, H.B. Park, J. Park, C. Seong Hwang, J.-C. Lee, S.-J. Oh, et al., Thermal annealing effects on the structural and electrical properties of HfO_2/Al_2O_3 gate dielectric stacks grown by atomic layer deposition on Si substrate, J. Appl. Phys. 94 (2003) 2563–2571.

[30] R.J. Potter, P.A. Marshall, P.R. Chalker, S. Taylor, A.C. Jones, T.C.Q. Noakes, et al., Characterization of hafnium aluminate gate dielectrics deposited by liquid injection metalorganic chemical vapor deposition, Appl. Phys. Lett. 84 (2004) 4119–4121.

[31] E.D. Litta, P.-E. Hellstrom, M. Ostling, Mobility enhancement by integration of TmSiO IL in 0.65nm EOT high-k/metal gate MOSFETs, Venice, Italy, Proc. Eur. Solid-State Device Res. Conf. (ESSDERC) (2013) 155.

[32] L. Chang, Y.-K. Choi, D. Ha, S.X. Bokor, C. Hu, T.-J. King, Extremely scaled silicon nano-CMOS devices, Proc. IEEE 91 (2003).

[33] C.M. Osburn, K.R. Bellur, Low parasitic resistance contacts for scaled ULSI devices, Thin Solid Films 332 (1998) 428.

[34] J. Seger, T. Jarmar, Z.-B. Zhang, H.H. Radamson, F. Ericson, U. Smith, S.-L. Zhang, Morphological instability of $NiSi_{1-u}Ge_u$ on single-crystal and polycrystalline $Si_{1-x}Ge_x$, J. Appl. Phys. 96 (2004) 1919.

[35] Paul R. Bessera, Simon Chana, Eric Patona, Thorsten Kammlera, David Browna, Paul Kinga and Laura Pressleya, "Silicides for the 65 nm Technology Node", 2003 MRS Spring Meeting, vol. 766, 2003.

[36] A. Lauwers, M. de Potter, O. Chamirian, R. Lindsay, C. Demeurisse, C. Vrancken, K. Maex, Silicides for the 100-nm node and beyond: Co-silicide, Co(Ni)-silicide and Ni-silicide, Microelectron. Eng. 64 (2002) 131.

[37] K. Bernstein, N.J. Rohrer, SOI Circuit Design Concepts, Kluwer Academic Publishers, Dordrecht, the Netherlands, 2002. Available online: <http://www.kluweronline.com>.

[38] G. James, S. Joseph, V. Mathew, The influence of metal gate workfunction on short channel effects in atomic-layer doped DG MOSFETs, J. Electron. Devices 8 (2010) 310–319.

[39] R.A. Soref, E. Cortesi, F. Namavar, L. Friedman, Vertically integrated silicon on insulator waveguides, IEEE Photonics Technol. Lett. 3 (1991) 22.

[40] A. Reznicek S.W. Bedell, H.J. Hovel, K.E. Fogel, J.A. Ott, R. Mitchell, D.K. Sadana, 300-mm SGOI/strain-Si for highperformance CMOS, in: Proc. IEEE Int. SOI Conf., Honolulu, HI, 2004, pp. 37–38.

[41] M.H. Liao C.Y. Yu, C.F. Huang, C.H. Lin, C.-J. Lee, M.H. Yu, et al., 2 μm emission from Si/Ge heterojunction LED and up to 1.55 μm detection by GOI detector with strain-enhanced features, IEDM, 2005.

[42] S. Kubicek, K. De Meyer, CMOS scaling to 25 nm gate lengths, in Advanced Semiconductor Devices and Microsystems, 2002, p. 259.

[43] J.-W. Yang, J.G. Fossum, On the feasibility of nanoscale triple-gate CMOS transistors, IEEE Trans. Electron. Devices 52 (2005) 5243–5275, pp. 1159–1164.

[44] L. Thylén, S. He, L. Wosinski, D. Dai, Moore's law for photonic integrated circuits, J. Zhejiang Univ. Sci. 7(12) (2006) 1961–1967.

[45] M.K. Smit, C. van Dam, PHASAR-based WDM-devices: principles, design and applications, IEEE Sel. Top. Quantum Electron. 2(2) (1996) 236–250.

[46] L. Thylen, Journal of Zhejiang University, Science 7(12) (2006) 1961–1964.

[47] L. Thylen, P. Holmstrom, L. Wosinski, B. Jaskorzynska, M. Naruse, T. Kawazoe, et al., Nanophotonics for low-power switches, in: I.P. Kaminow, T. Li, A.E. Willner (Eds.), Optical Fiber Telecommunications VI, Elsevier Science and Technology Books, Oxford, UK, 2013.

CHAPTER 5

Complementing Silicon With Other Materials for Light Emission, Efficient Light Modulation and Subwavelength Light Confinement

PART ONE: LIGHT-EMITTING SOURCES IN SI AS PHOTONIC MATERIAL

Since many lasers and optical amplifiers operate at standard telecommunication wavelengths of 1.3−1.5 μm, many research laboratories try to find novel materials and structures to be applicable for industry. The ideal vision is to have an optical source and amplifier that can operate at the same wavelength as the optical fibers. Hence, the best way to resolve this issue is to try to make a proper compound out of silicon-based materials.

Early attempts in this research field have shown that doping silicon with rare earth metals (lanthanides) may result in an excellent lasing material. The role of rare earth metals is not only limited to telecommunication wavelength lasing application and many studies are going

Monolithic Nanoscale Photonics—Electronics Integration in Silicon and Other Group IV Elements.
DOI: http://dx.doi.org/10.1016/B978-0-12-419975-0.00005-2
© 2015 Elsevier Ltd. All rights reserved.

on to utilize them in displays, optical amplifiers [1], light-emitting diodes [2], data storage [3], laser technology [1], etc.

Study of the electron configuration of lanthanides shows that not only their $4f$ orbital is partially filled but also owing to the presence of $5s^2$ and $5p^6$ electrons, the existing electrons in their $4f$ shell is shielded from external fields [4]. This means that the energy levels of these elements are largely insensitive to the host crystal in which they are placed. These elements exist as $3+$ and often $2+$ ions while incorporated in their hosts. Meanwhile, all the $3+$ ions show a high-intensity narrow-shaped band of intra-$4f$ luminescence with long luminescence lifetimes in the order of milliseconds. Since the $4f$ shell is shielded by 5s and 5p shells, the transitions in their solid host would be like free ions with a weak electron—phonon coupling and the dominant force affecting them is the one caused by spin—orbit interactions rather than applied crystal field. However, the crystal field interactions cause mixing opposite parity wavefunctions and allow partial intra-$4f$ interactions.

Selecting an appropriate ion with regard to the host can lead to intense and narrow-band emission across the visible region and into the NIR. Energy level diagrams for the isolated $3+$ lanthanide ions with partially filled $4f$ orbitals are presented in Figure 5.1. Excitation pathways for rare earth luminescence in solid hosts can be classified into two broad categories as direct and indirect mechanisms. In the case of direct mechanism, the following might be present: resonant optical excitation by the interaction of photons of appropriate wavelengths with specific rare earth $4f$ absorption bands, cathodoluminescence, and electroluminescence in semiconductor hosts involving hot electron collision with rare earth centers. Indirect mechanisms include carrier-mediated excitation transfer in semiconductors. As previously mentioned, rare earth metals due to their well-defined energy levels of the $4f$ shell electronic configurations have the ability to go under transition from the first excited state to the ground state $(^4I_{13/2} \rightarrow {}^4I_{15/2})$ at a wavelength, which is being employed to provide the gain in optical fiber amplifiers in long-distance telecommunication links worldwide [5]. The symbols representing the ground and excited states show how energy levels in rare earth metals are conventionally labeled. The letters (I, f) refer to the total angular momentum of the orbital of the ion. The left number (superscript) is the number of possible orientations of the total spin of the ion and the number to the right (subscript) is the total angular momentum of the ion.

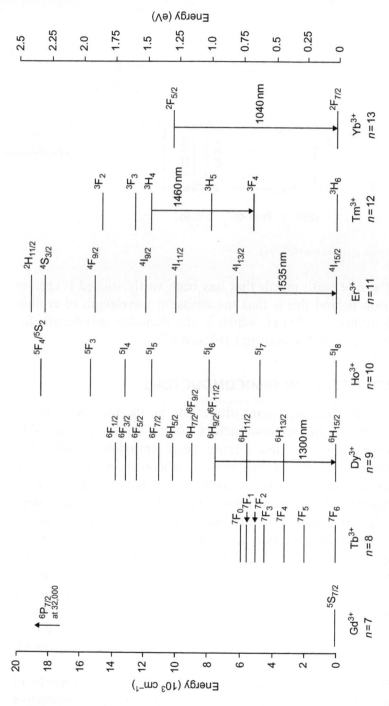

Figure 5.1 Energy levels of triply charged lanthanides. Only the important transitions are shown [5].

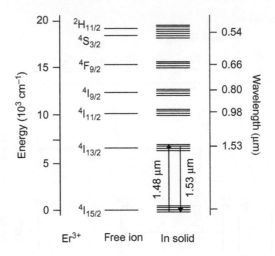

Figure 5.2 Energy levels of erbium ions (Er^{3+}) [6].

One of the rare earth metals that has been vastly studied is erbium and the reason behind this is that the emission wavelength of erbium ions (Er^{3+}) occurs at 1.53 µm, which is the desirable wavelength for existing telecommunication industry (Figure 5.2).

RARE EARTH METALS IN SEMICONDUCTORS

The present topic becomes more interesting as the incorporation of rare earth metals into semiconductors comes with the possibility of integrating a narrow- and intense band emission directly into microelectronic devices. For this reason, silicon as a host material has attracted attention as main interest since it is the semiconductor used for majority of microelectronics. Several techniques have been used in order to incorporate erbium ions into crystalline silicon (Si:Er) such as ion implantation and MBE. As previously mentioned, the lifetime of excited Er^{3+} is in the order of milliseconds and therefore the absorption and emission cross-section is quite small. In order to get a reasonable optical gain in amplifiers, high concentrations of Er^{3+} should be incorporated ($10^{20} - 10^{21}$ Er/cm^2), which owing to lack of sufficient space between the ions, the electric dipole–dipole interactions of the ions can reduce the gain performance. Instead, erbium shows large absorption cross-section in crystalline silicon (around 10^{-12} cm^2). However, much of the advantage gained from such efficient excitation is lost, thanks to the strong nonradiative de-excitation pathways that predominate in this material at room temperature. The low emission efficiencies and strong temperature quenching present severe limitations on the exploitation of bulk silicon as a host material. The low solubility of the rare earth ions in silicon is principally due to a combination of mismatch in ionic radii between the 3 + ions and silicon, and the predominantly sp^3 bonding of the silicon host. There is a strong tendency for erbium ions to cluster together in precipitates that provide further nonradiative de-excitation pathways, and for this reason, only relatively modest concentrations of erbium can be incorporated in silicon. Much of the work in this area has therefore been concerned with overcoming these problems [6–8]. Both direct optical and carrier-mediated excitation mechanisms are observed in erbium-doped silicon. For optoelectronic applications, the latter is of key importance, though photoluminescence studies of the former can yield important information about the chemical environment of the erbium ion, as well as allowing measurement of absorption and emission cross-sections. The mechanism of electrical activation of erbium in silicon is complex, involving carrier generation, trapping at erbium-related trap levels, and Auger transfer [9].

More investigations [10] showed that the free carrier concentration is proportional to the reciprocal of the Er^{3+} ion luminescence decay time. So by decreasing the decay time to below the radiative lifetime of erbium, the threshold of carrier concentration can be calculated, which in case of silicon as the host material is 1×10^{15} cm^{-3} [9].

The first photoluminescence of erbium-doped crystalline silicon conducted at room temperature [11,12] through co-doping with oxygen. The oxygen will provide solvation shell around erbium ions and oxygen-rich erbium-doped silicon showed two orders of magnitude higher luminescence intensity than the ones with low concentrations. The optimum concentration of oxygen in order to optically activate erbium-doped silicon should be an order of magnitude higher than the concentration of erbium itself [13,14].

POROUS SILICON

The first light emission out of porous silicon was observed in the visible range at room temperature [15] using hydrofluoric acid–based solutions to electrochemically etch silicon [16]. Owing to the efficient excitation pathways present in porous silicon, many scientists have tried incorporating dopants such as erbium [17]. Another advantage of porous silicon is the ability to tune its bandgap by changing the etch conditions and consequently the porosity of the material. As a result of this, it would be possible to activate rare earth luminescence with luminescence energies higher than bandgap of silicon. In addition, quantum-confinement effects produce long carrier lifetime and high degree of localization in spite of bulk silicon and amorphous silicon, which increases the possibility of interaction between the carriers and rare earth ions. Many studies have been done on the origin of photoluminescence in porous silicon host [18,19]. The results concluded that it is the quantum size effects, which lead to optical transitions and luminescence in the visible range. The model used is called quantum-confinement model, which gains support from the wavelength dependence of the luminescence on porosity. Sham et al. [20] also concluded that quantum-confinement model seems to be the only viable explanation.

Other studies were conducted in order to see the effect of late TM complexes in silicon. One of these metals is copper [21].

The doping was performed through evaporation of copper films through vertical diffusion and a fast cooling process using oil bath was performed to gain the most efficient emission. They found out that the luminescence intensity of copper-induced lines depends strongly on the diffusion temperature and the quenching rate of the sample. Through optimizing and keeping the quenching rate constant, they found out that by increasing the diffusion temperature of copper (from 700°C to 1100°C), the luminescence strongly increases and they related this property to the high solubility of copper at elevated temperatures.

A limited number of studies have also been performed on titanium- and vanadium-doped silicon. The reason behind it is these elements have the least solubility among all other late TMs [22], which implies weak luminescence intensities. Vanadium due to its low diffusivity tends to form precipitates and complexes with other impurities in silicon host and it is suggested that high-temperature annealing is required for the formation of vanadium complexes [23].

Reeson [24] also studied the photoluminescence of iron-doped silicon. They fabricated beta phase iron-doped silicon ($\beta FeSi_2$), which is a semiconducting orthorhombic phase. In order to fabricate this semiconductor, iron ion implantation with energy of 200 KeV and a dose of 5×10^{15} Fe cm^{-2} as well as long annealing time below the phase transfer temperature (900°C) are needed. Photoluminescence of $\beta FeSi_2$ at low and high temperature, with and without silicon filters, were performed. The purpose of using silicon filters was to eliminate emission above silicon band edge, which consequently means that only $\beta FeSi_2$ was being directly excited.

Optical properties of ruthenium-doped silicon (Ru_2Si_3) films and single crystals were evaluated by Lenssen [25]. The Ru_2Si_3 films were produced according to the same group's previous study using MBE [26]. The authors claim that the conditions in order to use Ru_2Si_3 as a light emitter is difficult to meet (extremely pure and defect-free material). Plus their optical measurements showed that the absorption coefficient at the band edge is several orders of magnitude smaller than other semiconductors (Figure 5.3). This can be either due to low DOS or low oscillator strength, which in any case will lead to small radiative recombination rate and eventually deficiency in light emission.

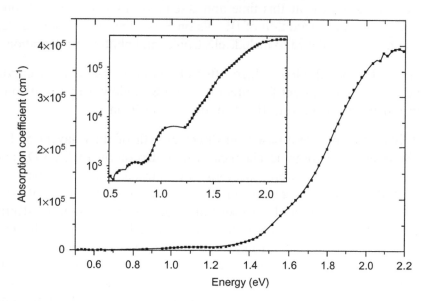

Figure 5.3. Absorption coefficient of a 140 nm thick Ru_2Si_3 film grown on silicon on sapphire substrate. The inset shows the same curve on a logarithmic scale [25].

PART TWO: COMPETING AND COMPLEMENTING TECHNOLOGIES AND MATERIALS TO AN ALL GROUP IV-BASED PHOTONICS APPROACH

III–V MATERIALS, PLASMONICS, AND ELECTROOPTIC POLYMERS (EOPs)

Introduction

As is obvious from the above, there are advantages as well as several shortcomings of silicon photonics, the most severe one of the latter category is doubtless the absence of a direct bandgap. Hence, it appears natural to complement silicon with other materials; the most prevalent and necessary candidates being III–Vs, mainly for light generation; EOPs for creating the electro-optic effect; metals for building nanophotonics structures. Such heterogeneous integration is facilitated by the general amenability of integrating these materials with silicon, albeit with some difficulty as for III–Vs.

Monolithic Integration of III–V Compounds on Silicon

The most efficient semiconductor light-emitting diodes and lasers are currently those based on III–V material, the prospects for group IV–based devices being unclear at this time and discussed in Chapter 3, part one. Hence monolithic integration of III–Vs on Si is a logical next step, a step that can also be widened to include electronics and photonics integration.

Thus, InP-based alloys which offer ubiquitously useful wavelengths have been a prime target for integration on Si. Below we present some alternatives for integrating III–V material on silicon.

Initial attempts were based on direct growth of thick layers of InP on Si using metal organic chemical vapor deposition (MOCVD) and later light sources were demonstrated using direct growth of InGaAsP/InP heterostructures on Si [27]. So far these have not been adopted by the industry due to several issues related to material degradation and early device aging caused by high lattice mismatch. This high lattice mismatch of around 8% and large difference ($\sim 50\%$) in thermal expansion coefficient of InP and Si motivated researchers to investigate other approaches for integration like hybrid integration using direct bonding [28,29] and adhesive bonding [30] of InP on Si. This technique has been the most successful method of integration and resulted in demonstration of light sources, amplifiers, modulators, and photodetectors on Si [31,32] with their performance comparable to those realized on planar InP substrates. Even though good results have been reported with this technique, truly monolithic approaches are desirable.

Recently some successful attempts to monolithically integrate III−Vs on Si have also been reported. Laser operation using III-dilute nitrides-based compounds like Ga(NAsP) lattice matched to Si substrates has shown encouraging results at 120 K, although room temperature operation remains to be shown [33]. Room temperature operation of InAs/GaAs quantum dot lasers on Si emitting at 1.3 μm has been realized [34] but so far emission at 1.55 μm has not been obtained. Another exciting approach to achieving heterogeneous integration of III−Vs on Si called epitaxial lateral overgrowth (ELOG) has shown promising results [35] and has the potential to monolithically integrate sources, detectors, and modulators for 1.55 μm although to date no such results have been demonstrated. InGaN-based light sources on GaN grown using ELOG have been demonstrated a long time ago, but to date no InGaAsP/InGaAs-based light sources on InP grown on silicon using ELOG have been reported.

In the ELOG technique, a thin film of the desired material is laterally overgrown on a dielectric mask selectively through defined openings from a predeposited seed layer on the host substrate. This dielectric mask hinders the infiltration of defects generated in InP seed layer due to its lattice mismatch with Si. In earlier studies on ELOG of InP on Si, even though the defects are stopped by the overlying dielectric mask, they are still observed above the openings [36].

Figure 5.4(a) presents a two-beam diffraction contrast dark-field TEM image showing ELOG InP on Si taken near the ⟨100⟩ direction. We observe that at point A, the defects are completely blocked by the SiO_2 mask, also within the opening. Nevertheless in this experiment, coalescence defects appear at point B [36].

Interface investigation of ELOG InP on SiO_2 by STEM (scanning tunneling electron microscopy) reveals that it is of high quality and the InP atomic lattice extends and retracts to conform to the curvature of the SiO_2 surface, thus ELOG appears as a promising tool for a truly monolithic integration of III−Vs and Si.

Plasmonics
Plasmonics has received much attention in recent years and has in fact exhibited an exponentially increasing stock of published work, only

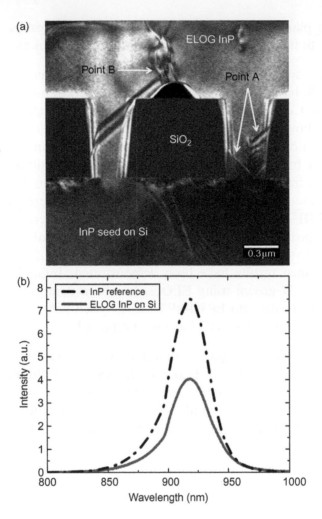

Figure 5.4 (a) TEM image of cross-section of sample. Point A indicates the blocked defects within the openings and point B indicates the defects generated due to coalescence of parallel growth fronts. The TEM images are high angular annular dark-field (HAADF) images taken along ⟨100⟩ direction. (b). Room temperature μ-photoluminescence measured from multiple openings on sample, indicating a decrease of the photoluminescence [36].

very recently starting to level off. This has been fueled by a general interest in nanotechnology, where negative dielectric constant or permittivity (epsilon, ε) materials, such as metals, offer the possibility of much higher concentration of light fields [37]. Figure 3.1 shows that the minimum lateral spatial field width for a planar silicon waveguide in air is ∼400 nm at a vacuum wavelength of 1550 nm. Thus, any attempts in nanophotonics should surpass this value, as measured in proportion to the relevant vacuum wavelength. Materials with negative ε, notably metals, offer a possibility for decreasing size and increasing the integration density in photonic light wave circuits in two ways: (i) By using the properties of materials with negative ε to concentrate light in ways other than employing total internal reflection as shown below and (ii) employing metal-based metamaterials to artificially generate very large effective media refractive indices, larger than those of silicon waveguides (metamaterials are artificial, man-made materials, where in general nanostructures are used to render novel macroscopic material properties, which the constituent materials do not exhibit in their natural appearance [38].

Both methods can allow true nanophotonics. A main and seemingly detrimental problem for many ICT applications has been the optical loss that is associated with both approaches.

A simple model for metal permittivity ε_m is the Drude model [39]:

$$\varepsilon_m = \varepsilon_\infty - \frac{\omega_p^2}{\omega^2 - i\tau^{-1}\omega} \tag{5.1}$$

where ε_∞ is the permittivity at wavelengths much shorter than the visible to NIR range considered here, ω is the light angular frequency, and τ is an electron scattering rate, responsible for optical losses. ω_p is the so-called plasma frequency:

$$\omega_p^2 = \frac{N_e e^2}{m^* \varepsilon_0} \tag{5.2}$$

Thus, the square of the plasma frequency is proportional to the free carrier concentration N_e and inversely proportional to the effective mass m^* of the freer carriers. Since one has to operate below the plasma frequency to get a negative dielectric constant ε, only materials with very high carrier concentrations such as metals can have a negative permittivity at optical frequencies.

A normal operating condition for plasmonics waveguides is that the operation frequency is lower than the plasma frequency as noted but larger than the inverse of the scattering rate τ.

Figure 5.5 shows the peculiar capacity of a metal/dielectric interface to concentrate and guide light along a single interface, in stark contrast

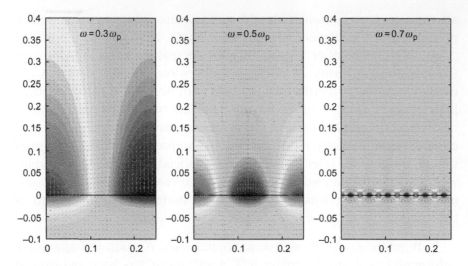

Figure 5.5 Optical field in a dielectric (above black line) and metal (below) interface, with light propagating left to right at the interface. When the light angular frequency approaches the so-called resonance frequency (the plasma frequency divided by the square root of 2 if the dielectric is vacuum) as in the picture furthest to the right, the field will be confined to infinitesimally small volumes and the group velocity goes to 0, in the absence of losses. This is in stark contrast to the guiding in the all dielectric waveguides relying on total internal reflection, as described above. Courtesy of Dr. Min Yan, KTH.

to the guiding in the all dielectric waveguides relying on total internal reflection, as described above. When the light angular frequency approaches the so-called resonance frequency for this structure (the plasma frequency divided by the square root of 2 if the dielectric is vacuum), the field will be confined to infinitesimally small extents in the vertical direction (far right in the figure) and the group velocity goes to 0, in the absence of losses, i.e., when $\tau^{-1} \to 0$.

However, a zero group velocity as obtained for the planar interface between a plasmonic medium and dielectric is not very appropriate for transporting information, if that is the application. One can, however, still utilize this maximal confinement and have a nonzero group velocity, by employing, for example, near field coupled metal nanoparticles. Each such metal nanoparticle is a resonator [37]; if they are near field coupled, the single resonance frequency is split into a band, very much like the case for optical and acoustical phonons, depending on polarization, and the dispersion curve now allows nonzero group velocity [40].

However, for existing materials, even this and other arrangements giving maximum field confinement and truly nanophotonics structures, on the order of several tens of nanometers, lead to horrendous propagation losses on the order of $1-10$ dB/μm. This effectively hinders all attempts to make more complex circuits and has been the main stumbling block for many ICT-type applications of plasmonics though not in other fields such as sensors, e.g., SERS (surface-enhanced Raman scattering) sensors. The great importance of a much lower loss negative epsilon material than existing ones, such as silver, could be amplified by paraphrasing Archimedes: *Give me a real, negative and practically implementable ε, over some optical wavelength range, and I will (perhaps) be able to use it as a massive leverage for integrated photonics.*

Electro-Optic Polymers

Modulation of the optical phase is the most versatile, since it can be the base of all types of modulation: amplitude (power), frequency, polarization.

The material of choice for high-speed optical modulators has for decades been lithium niobate (LN), which has been around since the 1970s. Though capable of performing clean phase modulation (without ancillary amplitude modulation), it has several disadvantages: limited electro-optic effect (conversion of electric field to refractive index change), limited overlap between the electro-optic material and the applied electric field, temperature sensitivity, cost issues in material, and fabrication. Thus other materials were sought after, and EOPs were one the alternatives.

EOPs have likewise been around for decades, early identified as potentially excellent materials for nonlinear optics such as second harmonic generation, as well as for optical modulators based on the electro-optic effect, the two effects are interlinked. In both cases, they could potentially show much superior performance to lithium niobate. Thus, the electro-optic coefficient r_{33} (see Eq. (2.15)) was generally more than a factor of 10 higher than in LN, with concomitant power dissipation saving of at least a factor of 100 in relation to LN, in reality higher due to more optimal device structures. However, a number of problems have plagued EOPs, the main one being stability: The material lost the electro-optic effect after times ranging from days to years. The reason is that in the virgin ate, the dipoles in the EOP are normally disordered, such that the material is centrosymmetric. Such a material has zero second-order tensors, thus the r_{33} and all other tensor elements are 0. Heating the EOP above the so-called glass transition

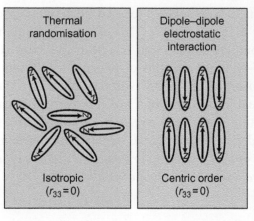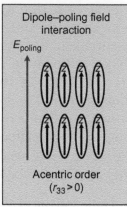

Figure 5.6 EOP dipoles in different states of order: Thermally randomized to the far left, ordered in a low energy but centrosymmetric state, with no electro-optic effect, (middle). Starting from the state to the left, with dipoles more or less free to reorient at an elevated temperature (above the glass transition point), a sufficiently strong electric field is applied, ordering the dipoles and a noncentrosymmetric structure exhibiting electro-optic effect results, far right. When the temperature is subsequently lowered, the structure is frown into place. The problem is that this is not an equilibrium state, a problem that has taken large efforts to solve. Courtesy of Dr R Palmer, Karlsruhe Institute of Technology.

temperature, T_g, makes it possible to align the disordered dipoles such that the material becomes noncentrosymmetric and thus exhibits an electro-optic effect. The temperature is then gradually lowered below T_g and the order frozen into the noncentrosymmetric structure. The problem was that this was not an equilibrium state, and the established order gradually disappeared. The poling process is illustrated in Figure 5.6.

Recently, however, breakthroughs have been accomplished in regarding EOP stability, among others by University of Washington, Seattle [41].

Thus, novel types of Nonlinear optical (NLO) guest–host polymer systems have provided remedy to the problem of combining high performance and stability, like chromphores with twisted motifs or charge transfer chromphores in dendritic encapsulation. Poled guest–host polymers exhibit electro-optic coefficient, r_{33}, values as high as 330 pm/V at 1310 nm, more than 10 times that of LN stated above.

It is anticipated that EOPs will change the entire landscape for a large class of photonics devices, since EOPs are Si compatible, much cheaper than LN to fabricate and capable of higher performance in terms of speed and energy dissipation.

AUTHORS' FINAL WORDS

The vision and challenge for group IV photonics and electronics are to integrate electronics and Group IV-based photonics. Here the Si-based photonics would be combined with Electrooptic polymers (EOPs), plasmonics (metals), III–Vs, and graphene or graphene-like materials in *"electronics-like" generic foundries, to mimic the process that took place for electronics in the 60s* and later. Silicon would here very much serve as a platform for photonics, with waveguide-based modulators, detectors, and WDM devices, and employing Si(GeSnC) alloys for increased performance.

However, most likely high-performance light sources will for the foreseeable future be implemented in hetero-integrated III–Vs, which can then also be used for modulation and detection.

For all high-speed, high-performance optical phase change modulation, especially involving advanced modulation formats, electrooptic polymers will most likely be used.

However, the real challenge resides in creating low cost, really nanoscaled, low-power dissipation truly integrated photonics electronics circuits for ubiquitous applications. Quantum leaps, such as those listed above, will be needed to achieve this grand goal.

For HEMTs, a SiGeSn template can be used for the growth of III–V materials. It is not so clear at this moment if the industry will go forward to GaN-based materials and neither for totally carbon-based technology. The requirement of several buffer layers prior to the GaN material makes it difficult to get accepted directly by the chip manufacturers. Any possible light source from GaN-based material will be outside the range of telecom wavelengths due to the wide bandgap of GaN but still useful for a wide range of rapidly increasing uses in other applications such as sensors and bio.

So far, there is a lack of a reliable technique to synthesize the carbon-based material (e.g., single or binary sheet of graphene) with high quality over an entire (300–450 mm) wafer. The other point related to graphene is that the manufactured detectors have low responsivity.

However, Ge may act as a material for both electronics and photonics. But the passivation, low solid solubility of dopants, low dopant activation, and poor thermal stability are issues which are considered as drawbacks for the use of this material in future. More research is required to improve the performance and the quality of the Ge materials. The bandgap engineering of group IV material has given rise to a monolithic solution for electronics and photonics where Sn alloys are proposed as direct bandgap material. GeSnSi alloys have excellent optical properties in terms of emission or detection of light with wavelengths of 1.55−2.2 μm but material quality is still an obstacle for integration. III−V compound materials are believed to continue to be dominant for laser applications, as noted, and probably for HEMT as well. Silicon and group IV photonics are here to stay, but the device and applications envelopes are presently somewhat unclear, the emergence of silicon photonics being by and large leveraged by the unparalleled enormous investments in R&D in silicon electronics.

The present ITRS will continue for at least another two decades, but it can be possible that photonic and electronic roadmaps will merge together in near future.

It is uncertain which type of new roadmaps will be created when the end of the current technology roadmap is reached. There is a tendency that the industry will manufacture 2D or 3D chips to increase

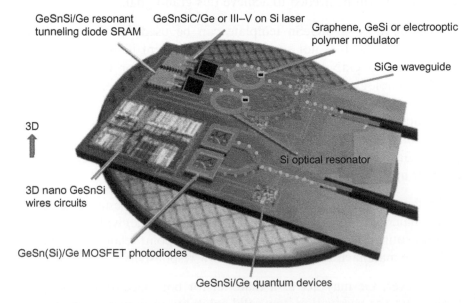

Figure 5.7 A 3D vision of photonic and electronic integrated components.

the number of transistors and follow a modified form of the today's "2D more Moore" to 3D beyond the Moore (Figure 5.7). Maybe new and advanced nanotechnology will play an even greater role.

REFERENCES

[1] M.J.F. Digonnet, Rare Earth Doped Fiber Lasers and Amplifiers, Marcel Dekker Inc, New York, 1993 ISBN:0-82479458-4.

[2] Pl for Er-doped silicon H. Ennen, 1.54 m electroluminescence of erbiumdoped silicon grown by molecular beam epitaxy, Appl. Phys. Lett. 46 (1985) 381.

[3] Ref data storage A. Loni, Relationship between storage media and blue photoluminescence for oxidized porous Silicon, Appl. Phys. Lett. 71 (1997) 107.

[4] S. Hüffner, Optical Spectra of Transparent Rare-Earth Compounds, Academic, New York, NY, 1978.

[5] A.J. Kenyon, Recent developments in rare-earth doped materials for optoelectronics, Prog. Quantum Electron. 26 (2002) 225−284.

[6] A. Polman, Erbium implanted thin film photonic materials, J. Appl. Phys. 82 (1997) 1.

[7] F. Auzel, Towards rare-earth clustering control in doped glasses, Opt. Mater. 16 (2001) 93−103.

[8] J. Lægsgaard, Dissolution of rare-earth clusters in SiO_2 by Al codoping: a microscopic model, Phys. Rev. B 65 (2002) 174114.

[9] G. Franzó, Understanding and control of the erbium non-radiative de-excitation processes in silicon, J. Luminesc. 80 (1999) 19−28.

[10] A. Suchoki, Auger effect in the Mn^{2+} luminescence of CdF_2:(Mn,Y) crystals, Phys. Rev. B 39 (1989) 7905−7916.

[11] H. Ennen, 1.54-μm luminescence of erbium-implanted III−V semiconductors and silicon, Appl. Phys. Lett. 43 (1983) 943.

[12] J. Michel, Impurity enhancement of the 1.54-μm Er^{3+} luminescence in silicon, J. Appl. Phys. 70 (1991) 2667.

[13] A. Polman, Erbium in crystal silicon: optical activation, excitation, and concentration limits, J. Appl. Phys. 77 (1995) 1256.

[14] G.D. Gilliland, Photoluminescence spectroscopy of crystalline semiconductors, Mater. Sci. Eng. R. Rep. R18 (1997) 99−400.

[15] L.T. Canham, Silicon quantum wire array fabrication by electrochemical and chemical dissolution of wafers, Appl. Phys. Lett. 57 (1990) 1046.

[16] A. Loni, Electroluminescent porous silicon device with an external quantum efficiency greater than 0.1% under CW operation, Electron. Lett. 31 (1995) 1288−1289.

[17] R.T. Collins, Porous silicon: from luminescence to LEDs, Phys. Today 50 (1997) 24.

[18] A.G. Cullis, Visible light emission due to quantum size effects in highly porous crystalline silicon, Nature 353 (1991) 335−338.

[19] V. Lehmann, Porous silicon formation: a quantum wire effect, Appl. Phys. Lett. 58 (1990) 856−885.

[20] T.K. Sham, Origin of luminescence from porous silicon deduced by synchrotron-light-induced optical luminescence, Nature 363 (1993) 331−334.

[21] J. Weber, Optical properties of copper in silicon: excitons bound to isoelectronic copper pairs, Phys. Rev. B 25 (1982) 7688.

[22] H. Conzelmann, Photoluminescence of transition metal complexes in silicon, Appl. Phys. A 42 (1987) 1–18.

[23] H. Lemke, Eigensehaften der dotierungsniveaus von mangan und vanadium in silizium, Phys. Stat. Solidi A 64 (1981) 549.

[24] K.J. Reeson, Is there a future for semiconducting silicides? Microelectron. Eng. 50 (2000) 223–235.

[25] D. Lenssen, Electrical and optical characterization of semiconducting Ru$_2$Si$_3$ films and single crystals, J. Appl. Phys. 90 (2001) 3347.

[26] D. Lenssen, Molecular beam epitaxy of Ru$_2$Si$_3$ on silicon, Thin Solid Films 371 (2000) 66–71.

[27] M. Razeghi, M. Defour, R. Blondeau, F. Omnes, P. Maure, O. Acher, First cw operation of a Ga$_{0.25}$In$_{0.75}$As$_{0.5}$P$_{0.5}$InP laser on a silicon substrate, Appl. Phys. Lett. 53 (1988) 2389.

[28] H. Wada, T. Kamijoh, Room-temperature CW operation of InGaAsP lasers on Si fabricated by wafer bonding, IEEE Photonics Technol. Lett. 8(2) (1996).

[29] A.R. Hawkins, T.E. Reynolds, D.R. England, D.I. Babic, M.J. Mondry, K. Streubel, et al., Silicon heterointerface photodetector, Appl. Phys. Lett. 68 (1996) 3692.

[30] G. Roelkens, J. Brouckaert, D. Van Thourhout, R. Baets, R. Nötzel, M. Smit, Adhesive bonding of InP/InGaAsP dies to processed silicon-on-insulator wafers using DVS-bis-benzocyclobutene, J. Electrochem. Soc. 153 (2006) G1015–G1019.

[31] A.W. Fang, R. Jones, H. Park, O. Cohen, O. Raday, M.J. Paniccia, et al., Integrated AlGaInAs-silicon evanescent racetrack laser and photodetector, Opt. Express 15 (2007) 2315.

[32] J. Hofrichter, O. Raz, A.L. Porta, T. Morf, P. Mechet, G. Morthier, et al., A low-power high-speed InP microdisk modulator heterogeneously integrated on a SOI waveguide, Opt. Express 20 (2012) 9693.

[33] S. Liebich, M. Zimprich, A. Beyer, C. Lange, D.J. Franzbach, S. Chatterjee, et al., Laser operation of Ga(NAsP) lattice-matched to (001) silicon substrate, Appl. Phys. Lett. 99 (2011) 071109.

[34] T. Wang, H. Liu, A. Lee, F. Pozzi, A. Seeds, 1.3-µm InAs/GaAs quantum-dot lasers mono-lithically grown on Si substrates, Opt. Express 19 (2011) 11381–11386.

[35] H. Marchand, X.H. Wu, J.P. Ibbetson, P.T. Fini, P. Kozodoye, Microstructure of GaN lat-erally overgrown by metalorganic chemical vapor deposition, Appl. Phys. Lett. 73 (1998) 747.

[36] H. Kataria, W. Metaferia, C. Junesand, C. Zhang, N. Julian, J.E. Bowers, et al., Simple epitaxial lateral overgrowth process as a strategy for photonic integration on silicon, IEEE J. Sel. Top. Quantum Electron. 20 (2014) 8201407.

[37] S.A. Maier, Plasmonics, Fundamentals and Applications, Springer, New York, NY, 2007.

[38] N. Engheta, R. Ziolkowaki (Eds.), Metamaterials, Physics and Engineering Explorations, Wiley-Interscience, USA, 2006.

[39] P. Drude, Zur Elektronentheorie der Metalle, Annalen der Physik 306 (1900) 566–613.

[40] W.H. Weber, G.W. Ford, Propagation of optical excitations by dipolar interactions in metal nanoparticle chains, Phys. Rev. B 70 (2004) 125429.

[41] L.R. Dalton, P.A. Sullivan, D.H. Bale, Electric field poled organic electro-optic materials: state of the art and future prospects, Chem. Rev. 110 (2010) 25–55.

Printed and bound by CPI Group (UK) Ltd, Croydon, CR0 4YY

03/10/2024

01040427-0016